CONTENTS.

PREFACE.

Farmers in the thickly populated Eastern and Middle States, or, in fact, east of the Mississippi River, cannot grow grains nor fatten beeves with the same profit as before the opening of the great West. Dairying still returns fair profits, but there is a widespread demand for cash crops adapted to farm culture, especially where railroads furnish quick access to towns and cities.

In response to this demand, we beg to offer a short list of farm vegetables that can be grown with greater profit than grain, with hints about growing them.

There is no real line dividing the vegetables of the market garden from those of the farm garden, but it may be assumed in a somewhat arbitrary way that those which do not yield at the gross rate of $250 per acre per year will not pay for the intense culture of high-priced land, although they will pay handsome profits in broad-acred operations under horse culture.

Before offering a list of money crops to farmers we shall have a word to say in the following pages about economic manuring. Larger cash receipts and smaller cash expenditures will result in better bank balances.

CHAPTER I.

MAKING THE SOIL RICH.

Everybody understands that the soil becomes impoverished by continued cropping, if no return be made in the form of manure or fertilizer. This impoverishment is sometimes real, while sometimes it is more apparent than real, owing to the exhaustion of only one or two elements of fertility.

Farmers have learned a great deal about agricultural chemistry since the introduction of artificial fertilizers. They know that while plants demand many things for their growth, there are but three elements which are in danger of being exhausted in ordinary cropping. These three things are nitrogen, phosphoric acid and potash.

Lime.—Lime is used on the land not for its direct results as a fertilizer, but because it has the ability to break up combinations already existing in the soil and set free the plant food that previously was in an insoluble form. Lime sometimes produces almost marvelous results; at other times no visible effects whatever. Hence, it is not a fertilizer, though in actual practice it is sometimes a fertilizing agent of great value. Land that has been much manured or long in sod is likely to be benefited by lime.

Artificial manures, on the other hand, furnish real plant food in soluble form, and may be expected to produce crops invariably, year after year, if the soil be sufficiently moist. When a fertilizer contains nitrogen, phosphoric acid and potash it is said to be complete. When any element is missing the fertilizer is said to be incomplete. Ground bone, wood ashes, South Carolina rock, kainit, etc., are examples of incomplete fertilizers.

Barnyard Manure.—Barnyard manure is the best of all known fertilizers. Not only is it complete in character, but it has the highly valuable property of bulk. It is rich in humus or humus-forming materials. It opens and ventilates the soil, and improves its mechanical condition to a remarkable degree. Humus is a name for decaying organic matter.

American market gardeners deem it entirely safe to use fifty to seventy-five tons of barnyard manure to the acre of ground in their intensive cultural operations. American farmers seldom apply more than ten or fifteen tons of such manure to the acre in the open field.

The manufacture of artificial fertilizers had its origin in the fact that cultivators could not get enough manure from natural sources, and, hence, were compelled to go into the market and buy nitrogen, phosphoric acid and potash in other forms.

Closer Economy.—With the increase of competition and consequent fall of prices a closer economy in cost of production is necessary. Prices have fallen most in respect to commodities that will bear long-distance freight transit and less in respect to the more perishable products of the soil. Hence, farmers have widely turned attention to small fruits and vegetables for money crops, instead of grains, and are now studying how to fertilize these crops in the most effective and economical manner.

It is very evident that while great quantities of fertility are demanded by the new crops, there is no such margin of profit in their culture as to warrant wasteful methods, and no losses of home-produced fertility can be tolerated.

As to Saving Manure.—A penny saved is a penny earned. A half ton of manure saved is a dollar earned; and, conversely, a half ton of manure wasted is a dollar wasted.

In many American barnyards much of the manure is lost, partly by leaching and partly by escape of ammonia. It is estimated that as much as a third of the natural manure produced in this country is practically thrown away.

The Cornell Station has announced that a pile of horse manure exposed to the weather will lose half of its value in six months. The Kansas Station reaches nearly the same conclusion about farmyard manure.

Manure stored under cover may lose from 14 to 30 per cent. of its nitrogen (ammonia); and as this element is the most expensive of all to buy, it is evident that the loss is a very serious one, and one that should be avoided if possible.

General Principles of Storage.—Having pointed out the fact that on many farms there is a loss of a large amount of excellent manure, it is now in order to name a remedy. The compass of this book is so limited that it is necessary to go straight to the point, omitting a detailed account of the chemical processes involved.

The best-known method of keeping all the manure produced by farm animals is storage under a closed shed, supplemented with chemical preservatives. The shed need not cover the barnyard, but merely the manure pile. The preservatives cost little money, and eventually go to the soil in the form of excellent fertilizers. Not a cent paid for them need be lost.

The manure shed should be large enough to work in with comfort; large enough to permit the heap or heaps of manure to be turned, worked over and shifted from place to place. A clay or earth floor will answer every purpose, and the shed may be of the cheapest character, provided it will turn the rain. The floor of the manure shed should slope inward from all directions, and the drainage around the shed should be outward, so that no rain-water or snow-water can enter.

In theory, it may be best to put fresh manure on the land as quickly as possible. All leaching is then received by the soil, and little is lost, except through the air.

In practice, this plan is not always a good one. It costs more to make ten trips to the field than one trip, and valuable time is wasted. It is quite out of the question to haul out manure every day or even every week. Besides, it is necessary in actual practice, especially in gardening or truck farming, to cover a whole piece of ground at one time, so that it may be plowed and seeded for the coming crop. The ground is usually available only a short time before this preparation, having, perhaps, been occupied by something else. It is desirable, moreover, that the manure when applied shall be ready for immediate service as plant food, which is not the case with the raw product. Fresh manure is but sparingly digestible by plant roots. Quicker cash results will be secured by applying prepared manure to the soil than by applying the product fresh from the stable.

The manure shed has already been mentioned. A few dollars will build it. Sometimes a half barrel is sunken in the centre of the manure shed, and the drainage from the manure heaps collected there, and returned to the tops of the heaps. It is occasionally necessary to add water, when turning manure, to secure the desired degree of dampness and a gentle fermentation. This fermentation will cause the litter to fall to pieces, and will convert it into quickly-available plant food.

No one who has never tried it will expect the generous heaps which will follow systematic and persevering efforts to accumulate and stack up the available manure materials on any farm.

Preservatives.—The best-known common preservatives of manure in storage are gypsum, kainit and acid phosphate.

Gypsum or land plaster holds ammonia, and is thus of the highest value as a preservative. Gypsum must be moist to be effective, and, hence, should be used regularly upon the fresh manure.

Kainit, which is a low-grade sulphate of potash, checks fermentation, and hence prevents loss of ammonia. It contains much salt, and attracts and holds moisture. It should not be used under the feet of animals.

Acid phosphate contains much gypsum, and unites with ammonia that would otherwise escape.

The Geneva (N. Y.) Station recommends the use of one of the following per day:

	Per Horse. Pounds.	Per Cow. Pounds.	Per Pig. Ounces.	Per Sheep. Ounces.
Gypsum	1½	1¾	4½	3½
Acid phosphate	1	1⅛	3	2½
Kainit	1⅛	1¼	4	3¼

The advantage of using kainit and acid phosphate are that they add potash and phosphoric acid respectively, in which barnyard manure is likely to be deficient. In some soils the potash will be preferable; in others, phosphoric acid will do more good.

Value of Manure of Each Kind of Animal.—It has been figured out that the average value of horse manure per year is $27 per animal; cattle, $19; hogs, $12; and sheep, $2. But these are not the only sources of manure on the farm. The hen-house will annually yield manure to the value of 25 to 50 cents per fowl, if intelligently cared for. The outhouse will produce fertility to the amount of $10 to $50 per year, according to the size of the family, the precautions as to loss by leaching, and the care given. The kitchen slops, including the scraps, are worth $10 to $25 per year, if properly composted. The wood ashes have a distinct and high fertilizing value; but not in the hen-house, where they are worse than wasted. And even coal ashes can be turned to account.

Professor Roberts has suggested $250 per year as a conservative estimate of the value of the manure produced during seven winter months on a farm carrying four horses, twenty cows, fifty sheep and ten pigs. The estimated value may be made much higher in cases where farmers are willing to use thought and labor in preparation and preservation of home-made manures.

Solid Manure and Liquid Manure.—The urine is the most valuable portion of the excretion of animals, according to the tables of the agricultural chemists. It is especially rich in nitrogen, and, hence, its strong odor under fermentation. It is also rich in potash. Its place is on the manure heap, not in a ditch leading to a brook. If it collects in quantities beyond the absorbing power of the manure pile, it should go on the compost heap or else be diluted and at once put upon the land.

When to Fertilize.—The land is a good bank in which to deposit money in the form of manure; but there are certain portions of the year when the land bank declares no dividends. It is safe to put manure upon an unfrozen soil at any time, but the best, the quickest, and the largest results are obtained by manuring during the growing season, preferably just before planting the crop. Small applications, often repeated, are preferable to large, though rare, applications. Plants, like animals, consume small amounts of food each day, and cannot take a year's food at a single meal.

Humus.—Humus, often referred to by agricultural writers, is a name for decaying organic matter in the soil. Green crops turned under, grass roots, stubble, leaves, long manure, etc., form humus. The term is a comprehensive one. Humus is a dark-colored substance, abundant in all rich ground. A lump of manure that has been lying in the ground for a year or two has become, practically, a mass of humus.

Minute Soil Workers.—In all good soils there are myriads of small organisms, whose duty is to destroy organic matter and convert it into soil, or into humus, or into plant food.

This explains the superiority of good, moist soil as compared to coal ashes for making compost heaps. Coal ashes are worth sifting, if the work can be done automatically; that is, by simply pouring the ashes upon a sloping wire screen. The coarse portion of the ashes, if not worth reburning, will at least make good walks, drives or road beds, while the fine portions make excellent absorbents to put under hen roosts.

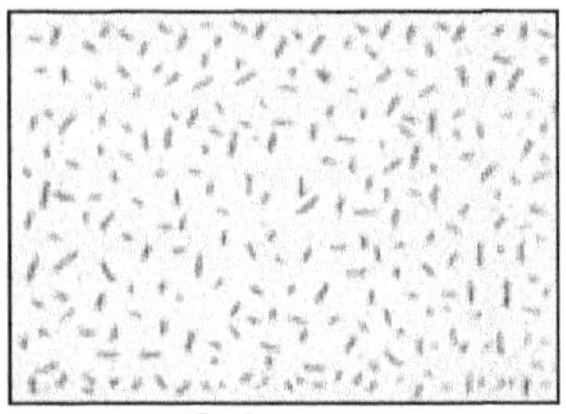

Some of the many Forms of Bacteria. (Magnified).

Hen manure and the product of the outhouse, whether containing sifted ashes or not, should go speedily into a heap of moist earth, for this earth will furnish the organisms to quickly convert the excreta into valuable soil. Sifted coal ashes usually contain some fertility on account of wood, garbage, etc., burned in the kitchen stove, but have value mainly as absorbents. Moist loam, on the other hand, teems with life, and has the wonderful ability not merely to hide organic matter, but to actually change its character, converting it into soil that retains none of its original characteristics. What was malodorous manure, offensive to smell and touch, is changed into an odorless, dark-colored material that leaves no stain upon the hands, and which is plant food of the best and most available character.

Economy in Manuring.—True economy in manuring demands a comprehension of these simple matters. The methods are inexpensive, and are within the reach of every tiller of the soil.

The whole matter may be summed up in a few words, as follows: Waste nothing, permit no fermentation or leaching, use preservatives, and learn the true art of making composts, including the functions of the minute organisms just described.

No better use can be made of rainy days in summer or winter than in caring for manure; turning the piles, making compact stacks, adding needed moisture and preservatives, shaking out all lumps and putting undecayed portions into the centre of the heap.

Ton after ton of the best kind of fertilizer can be accumulated on every farm in this manner, including not only what is now lost through careless handling, but also a large amount of good material that is now entirely overlooked on many farms. All rubbish, all litter, all dirt, has a fertilizing value. If certain waste products must go to the bonfire, the ashes can at least be saved and used during the next growing season. It is sometimes better to burn weeds and certain tough vines than to attempt to compost them; but the ashes should not be wasted. It is the saving of many little things that counts in the yearly total. Labor is money, but it is better to invest labor at home than to go to the fertilizer-maker for supplies and pay out cash.

The Fertilizer Man.—The fertilizer man will always be with us, because he has a true place in the economy of the farm and garden. We must go to him for the preservatives already mentioned—for gypsum, for kainit and for acid phosphate; and also for complete fertilizers. These articles are all comparatively cheap. The fertilizer man can make but modest profits upon them. The purchase of high-grade goods from well-known and honest makers is to be commended, for it is strictly economical.

The thing to be avoided is the blind buying of fertilizers from unknown or irresponsible makers or agents. This is worse than buying a cat in a bag, and results in great waste of good money.

Wood Ashes.—Wood ashes is rich in potash, and is particularly valuable with potatoes, fruits of all kinds, etc. But it is a great error to mix wood ashes with fresh manure of any kind, especially with hen manure, as the escape of ammonia is hastened and much value is lost.

Natural and Artificial Manures.—Where the home supply of manure is insufficient for a piece of ground, necessitating the addition of artificial manure, it is universally conceded to be good practice to stretch the natural product over the whole tract and then to go over the whole tract with an artificial fertilizer.

Irrigation. [A]—In connection with a review of the home sources of manure the item of irrigation must not be overlooked, for it is thoroughly well established that water is a carrier of appreciable amounts of fertilizing materials. In European countries large areas of pasture and mowing lands are fertilized by water alone, the irrigation being regarded as of great value on this account, aside from the fact that it supplies moisture to the grass roots. Most streams in the United States contain more or less sewage, and in respect to irrigation are valuable on that account.

Waste Products.—Many waste or by-products, available for use as fertilizers, come from time to time within the reach of the farmer or gardener, especially to those living near towns or railroads.

The average market house, be it said with regret, is none too clean, and refuse in considerable amounts could be had there for the trouble of sweeping. Cattle cars often contain several inches of valuable droppings, to be had for next to nothing. Street-scrapings are worth the trouble of hauling, [Pg 18] [Pg 19] if the distance is short. The manure lost on the highways is very great in amount, and may be worth the cost of collection. There will some day be a machine for gathering this manure from the roads by horse-power, as it would amply repay the expense of driving such a machine along every much-used highway.

An Illustration from Johnson & Stokes' New Book—"Irrigation by Cheap Modern Methods."

Value of Manure.—Dr. Beal figures the values per ton of the several farm-made manures as follows: Hen manure, $7.07; sheep, $3.30; pigs, $3.29; horses, $2.21; cows, $2.02.

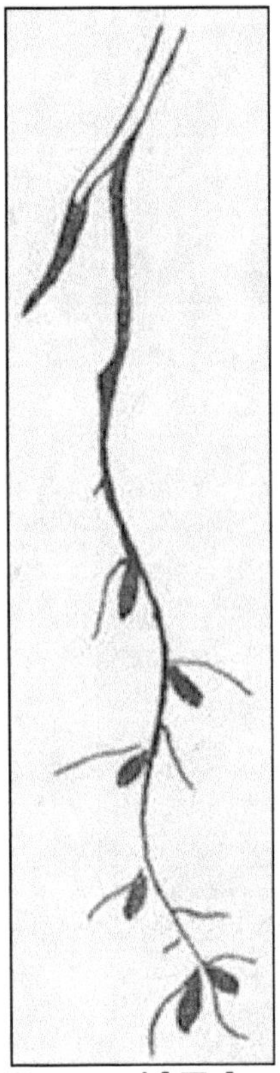

Bacteroid Tubercles on Red Clover Root. Drawn from Nature.

These figures are based on the assumption that the animals are well fed, and that no leaching of the manure is allowed, with gypsum used as a preservative, and good care exercised in all respects. It must not be supposed that all manure has such value, or that any manure will retain such value under careless treatment.

Green Manuring.—The system of green manuring, as formerly understood and practiced, had two purposes in view. One was to supply the soil with needed humus; the other to furnish winter protection and prevent washing. The practice is a very old one and has much to commend it. Not only do plant roots draw up fertility from considerable depths, to be afterward deposited in the superficial soil when the growing crop is turned down by the plow, but the process favors chemical changes in the soil by the admission of air and sunlight and by the decomposition of leaves, stems and roots. But nothing whatever in the way of new fertility is added by turning down a rye crop, for instance.

Crimson or Scarlet Clover, a Nitrogen Gatherer.

Cultivating the Legumes.—The present system of green manuring contemplates something in addition to what was formerly gained, for agricultural sciences now recognizes the fact that nitrogen, the most expensive element of fertility, can be taken from the air and added to the plant food in the soil by means of certain plants which have the peculiar habit of regularly forming little tubercles or lumps on their feeding roots. These lumps are to be found on plants in perfect health, and are not parasitical in any hostile sense. The lumps are filled with small living organisms called bacteria, and, hence, have been called bacteroid tubercles. The minute tenants slowly but surely secrete nitrogen, and put it in a form adapted to plant growth.

The plants which bear these root lumps belong to a group called legumes, of which clover, peas, beans, vetches, etc., are familiar examples.

Curiously enough, nearly all the leguminous plants are thus fitted by nature by means of the root lumps to act as soil enrichers, and these plants have, therefore, assumed the highest agricultural significance.

It is well known that such crops as cowpeas, crimson or scarlet clover, common red and pea vine or sapling clover, Soja beans, vetches, etc., can be used to add nitrogen to the soil in commercial quantities. The gain of new material, expressed in money, has been estimated as high as $25 per acre. This, therefore, is the avenue through which the farmer can most economically supply nitrogen to his land. If he will exercise all the economy heretofore suggested in the care of natural manures, and will grow legumes, he will not have much occasion to buy nitrogen in the market.

Grass vs. Clover.—An idea of the great fertilizing value of the leguminous plants as compared with grasses may be obtained by a study of the following analyses from U. S. Farmers' Bulletin No. 16, by Dr. E. W. Allen, on "Leguminous Plants for Green Manuring and for Feeding":

Hay from	Assumed Yield. Per Acre. Tons.	Fertilizing Value in Crop. Per Acre. Nitrogen. Pounds.	Phos. Acid. Pounds.	Potash. Pounds.
Red top (a grass)	2	23·0	7·2	20·4
Timothy (a grass)	2	25·2	10·6	18·0
Red Clover (a legume)	3	62·1	11·4	66·0
Alfalfa (a legume)	3	65·7	15·3	50·4
Cowpea (a legume)	3	58·5	15·6	44·1
Soja bean (a legume)	3	69·6	20·1	32·4

Nitrogen, Phosphoric Acid, Potash.—We have just noted the cheapest source of nitrogen. It can be collected by root tubercles at less than the commercial rate of 14 to 17 cents per pound.

Phosphoric acid can be best secured, if a new supply becomes necessary, in the form of ground bone or in the form of acid phosphate. Either of these articles, if bought from a reliable dealer, is a good and economical thing to use.

Potash is to be had most cheaply, perhaps, in the manner suggested heretofore: by the use of kainit as a preserver of stable manure. The kainit performs a double purpose if used in that way, and thus gets upon the land in a cheap manner.

Muriate of potash and sulphate of potash are high-priced articles, but when bought from good houses are fully worth the money they cost. Except for the use of kainit, just mentioned, the muriate or sulphate would be the more economical form.

Potash or phosphoric acid (or both), as may be determined by circumstances, are needed to aid crimson clover in its growth, and with the clover form a perfect manure.

Barnyard manure is a perfect fertilizer, especially when preserved with kainit or acid phosphate; and a leguminous crop, if stimulated with phosphoric acid and potash, leaves the land in fine cropping condition.

Value of Green Manures.—The cash value of green manuring is somewhat a matter of location. On light, sandy soils it will be found wise to turn the whole crop under with the plow, while on heavy loams this plan is of doubtful benefit. On the latter land it is conceded to be better practice to harvest the crop and feed it to stock, and return the resulting manure to the land.

Maximum Amounts of Manures.—Nobody has yet ventured to fix the maximum amounts of natural or artificial manures that soils will bear, but these amounts are great. Reference has already been made to the number of tons of stable manure per acre used respectively by market gardeners and farmers in America. As to commercial fertilizers, the quantity has been pushed up to two tons per acre, with enormous crops in consequence, and with no bad results where the constituent of the fertilizer were well balanced and where the water-supply was ample. It is quite easy, however, to scorch or burn the foliage of growing plants by the improper use of acid fertilizers in dry weather. Of course, no such amount as two tons per acre would be used in ordinary farming or farm gardening, but only in certain intense cultural operations.

CHAPTER II.

CHOICE OF LOCATION.

Almost every farm has a choice spot for a garden, some favored location where the soil is warm and mellow, and where, perhaps, shelter is afforded by hill or woodland. Such a spot, especially if it can be artificially irrigated, is capable of great things in the way of growing truck.

The place of all others, if it can be had, is a rich meadow bank, on ground low enough for gravity irrigation and yet high enough to be out of the way of floods. Such a location is by no means rare. There are countless acres fulfilling these conditions, and every acre thus situated is capable of yielding in vegetables twenty-fold its value as pasturage.

Such a meadow needs a few lines of underdrains and an irrigating ditch along the highest feasible level. Deep plowing of low land will rarely bring up the sub-soil, and, after a good coat of lime, the application of manure may be carried to almost any extent, with good results assured in advance.

If a meadow is not available, the farm gardener will do the next best thing, whatever that may be, in choosing a place for vegetables, trusting the rainfall and depending on manure and good tillage for satisfactory crops.

As to Growing.—The one point to be emphasized about the production of truck for market is that quick growth is necessary for quality, and, hence, for profits. Good soil, good cultivation and sufficient moisture are the essentials for rapid growth.

As to Marketing.—A point of prime importance for all [Pg 25] [Pg 26] producers to remember is that price is largely a matter of taste and fancy. If the consumer can be attracted by the good appearance of vegetables or fruit, a sale is certain to be made. It will pay handsomely to keep at home all medium or second-quality stuff, offering nothing but the best for sale.

Reproduction of a Photograph taken in Dock Street Wholesale Market, Philadelphia.

In the great wholesale and retail markets of Philadelphia, New York and Boston good stuff always moves quickly at fair prices, while poor stuff begs for buyers at rates yielding no profit to anybody. The wholesaler is frequently blamed for failure to obtain good prices when the fault is really with the producer, and is chargeable to poor stuff or poor packing.

There is a good business opening everywhere for truckers who will ship only first-class stuff in new packages. Such produce reaches what is known as the fancy trade, and there is more than a living in it for enterprising growers.

Truckers who rush their stuff to market in an unwashed, unsorted condition, in old or unclean baskets or boxes, may make expenses out of the business, but they will never do much more. There is a premium on quality and appearance.

CHAPTER III.

VEGETABLES SUITED TO FARM CULTURE EVERYWHERE.

Loading the Market Wagon.

In this chapter are grouped a number of vegetables of easy culture. They may be grown with success almost anywhere. Some of them are produced by market gardeners, but by reason of the amount of ground which they occupy they are more particularly adapted to horse culture by farmers.

The chapter will treat briefly of asparagus, beans, beets, cabbage, carrots, sweet corn, horseradish, parsnip, potato, pumpkin and squash, salsify, tomato, turnip, etc.

ASPARAGUS.

Asparagus demands a deep, rich, well-drained soil. Its culture is profitable, and it yields ready cash at an early season of the year, when other sales are limited. The cutting term covers six weeks, beginning (at Philadelphia) in the middle or latter part of April. Cutting must here cease in June, in order to give the roots ample time to regain strength and make vigorous tops. The gross product per acre, near Philadelphia, expressed in money, is, perhaps, $200 at this time.

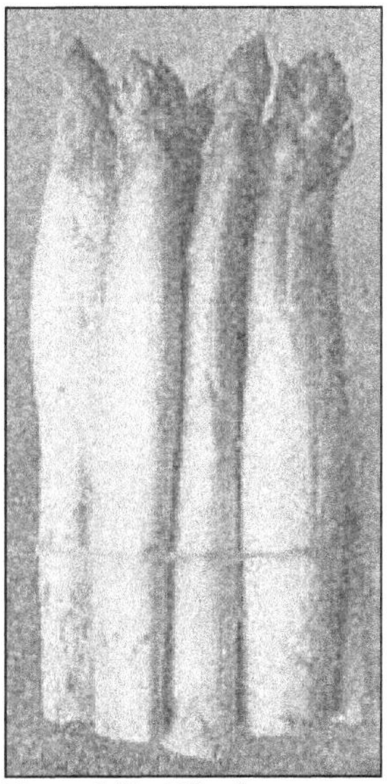

Donald's Elmira.

In selecting a situation for a bed, a warm spot should be chosen, having a deep and mellow soil, and with good natural or artificial drainage. A small area is better than a large one, as being more likely to receive sufficient manure; and it is desirable that the land should have been tilled for a year or two before the planting of the roots, and a heavy coat of manure incorporated with the soil—the more manure the better.

Roots.—The roots are set in early spring, in deep trenches, 5 or 6 feet apart, made with a plow. If the plow be run both ways and the loose dirt shovelled out, it is quite easy to reach a depth of 15 or more inches. It is not material whether strong one-year-old roots or two-year-old roots be used.

Varieties.—As to varieties, it is almost as much a matter of culture as of name; still, there are better and worse kinds. Asparagus varies in color from purple to green, and even to white. There are certain so-called mammoth sorts, whose shoots are larger, but less numerous than the old-fashioned kinds. There is a slight difference in flavor, also, but the preference of the local market must determine the farmer in making a choice of roots. If a green "grass" be preferred, that kind can be had from seedsmen; but, no matter how carefully the roots may be grown, there will be some slight variations in the color of the shoots, for asparagus does not always come true from seed. Market gardeners usually sort their asparagus shoots at bunching time; always for size, and sometimes for color, especially when supplying a fancy trade.

As a rule, it is wise to select a variety that will produce a good number of large-sized shoots, such as Donald's Elmira or Barr's Mammoth, and trust to manure and culture for the best results. Quality of shoots depends on quick growth, and size depends somewhat on distance of the root under ground. The deeper the asparagus root under the surface, the larger in diameter will be the shoot, provided the plants are not crowded; a fact of which gardeners often take advantage by heaping soil up over the crowns of the plants during the growing season.

Asparagus.—Donald's Elmira is one of the best for the North. Palmetto is the asparagus generally grown in the South. For descriptions, see "Johnson & Stokes' Garden and Farm Manual."

Setting out.—The young roots should be set carefully, crowns up, at intervals of 1½ to 2 feet, in the deep furrows or trenches heretofore mentioned. A few inches of manure can be put in the bottom, covering slightly with soil and about 6 inches of soil put upon them. The spaces between the rows may be cultivated during the first year, and some quick crop grown there. The working of this crop will gradually fill up the furrows about the stems of the young asparagus, which, during the first year, is quite small and insignificant in appearance. By fall, the furrows will be entirely filled and the surface of the patch level. The asparagus slug, the larva of the well-known beetle, may be kept down by occasionally dusting with slacked lime containing Paris green. The following year the asparagus will show up to some advantage, but should not be cut. The third year (second after planting) will yield some marketable shoots; but cutting should not continue more than two or three weeks. The fourth year the bed may be said to be in full bearing.

Treatment.—The spring treatment of an asparagus bed in profit begins with a light plowing parallel with the rows, great care being observed to use a wheel on the plow so that not more than a few inches of soil may be turned, lest the crowns be cut and injured. The bed then lies until the cutting season is well advanced, when the plow may be again used.

The first plowing was merely to break the surface of the ground and turn under the winter coat of manure, leaving the land level. The second plowing

Varieties and Types.—The varieties of beans are well-nigh endless. Some demand poles, while some are dwarf, being called bush beans. The influence of man has developed the bean into a vast number of different forms, which frequently show a disposition to revert or go back to some ancestral type, no matter how carefully the seeds may be kept.

The pole beans, in general terms, yield larger crops and bear through a longer season than the bush beans. The green-podded beans, as a rule, are more prolific and more hardy than the yellow-podded or wax beans. The climbers demand a whole season, and bear until frost. The bush beans are mostly employed where two or more crops are demanded per year from the ground.

The so-called cut-short or snap-short beans are those in which the whole pod, in its green state, is used for food. They are of both types, climbing and bush. The Lima forms include a number of distinct beans, differing greatly in size and shape and also in habit of growth.

BUSH BEANS (green pod).—We recommend Improved Round Pod Extra Early Valentine; also, New Giant Stringless Valentine.

BUSH BEANS (yellow pod).—Wardwell's Kidney Wax and Davis' White Wax are largely grown in the South for shipment North. Valentine Wax is recommended for the North. For descriptions, see "Johnson & Stokes' Garden and Farm Manual."

WHITE FIELD OR SOUP BEANS.—We recommend Day's Leafless Medium and New Snowflake Field. For descriptions of these and other varieties, see "Johnson & Stokes' Garden and Farm Manual."

POLE LIMA BEANS.—We especially recommend Ford's Mammoth Podded Lima and Siebert's Early Lima.

POLE SNAP BEANS.—Golden Andalusia Wax is one of the best yellow-pod pole beans, and Lazy Wife's one of the best green-pod sorts.

DWARF LIMA.—Dreer's, Burpee's and Henderson's represent three distinct types.

For full descriptions of beans, see "Johnson & Stokes' Garden and Farm Manual."

Location.—In choosing a spot for bean culture the farm gardener should select good mellow soil that has been manured the previous year. Fresh manure produces an excessive growth of vine at the expense of pods.

Making Ready.—Much stable manure, which is rich in nitrogen, should be avoided. In good ordinary soil, with some rotted manure from the previous crop, the bean plant will do well. It will obtain nitrogen, in great part, from the air, as already explained. Old manure is very favorable as a starter, as it contains the minute organisms mentioned in the preceding pages. Complete fertilizers or those containing phosphoric acid and potash must be supplied. Only nitrogen is derived from the air.

Soil Inoculation.—The soil of a new bean patch is sometimes inoculated with soil from an old patch, to get quick action of the bacteria (little organisms), which form the lumps or tubercles on the roots. The scattering of a little soil over the surface is all that is required.

Care should be taken to avoid the transfer of soil for this purpose from a patch affected with rust or blight, as diseases are carried from place to place with only too much ease.

When to Plant.—Beans may safely be planted when the apple is in bloom, in May; not so early as peas, as beans are less hardy. The ground should be dry and warm. Beans of all kinds demand shallow planting, as the seeds must be lifted from the ground in the earliest process of growth. The seed swells, bursts, sends a shoot (radicle) downward, and the two parts of the seed, called the seed-leaves, are pushed up into the daylight. Small round beans can take care of themselves, as they turn easily in the soil, but lima beans often perish in the effort to get above ground. This is why lima beans should always be planted eye down, and less than an inch deep. A half inch is deep enough for most beans. If lima beans are wanted extra early, they should be started on small squares of inverted sod, under glass.

The earliest bush beans yield marketable pods within forty to fifty days from planting; the pole beans in from seventy to ninety days from planting. There should be successional plantings made of the bush beans from the first date to within fifty days of frost. The different types of beans are fully and carefully described in the seed catalogues.

Distances.—Poles for beans should be set about 4 feet apart each way; or, in single rows, about 3 feet apart. Not more than three or four plants should be allowed to a hill. Wires stretched between posts, with strings down to the ground, are sometimes used. The bush beans are planted in rows 3 feet apart for horse culture, or half that distance where a hoe or hand cultivator is to be used. The plants in the rows should stand 3 or 4 inches apart for best yield.

Plant of the New Valentine Wax Bean. The Earliest Wax or Yellow Podded Snap Short.

On a Large Scale.—In large field operations, where the dried bean is the object in view, a clover sod is a favorite location. The ground is enriched by 400 or 500 pounds of complete fertilizer, and the beans are planted with a grain drill, using every fourth tube. The culture is by horse-power, and the vines are pulled by hand or by means of a bean-harvester, and threshed with a flail or grain thresher. These white grocery beans are sold everywhere in large quantities.

Cultivation.—All bean cultivation should be shallow. Nothing is gained by cutting the feeding roots. The climbing sorts twine "against the sun;" that is, in a contrary direction to the apparent motion of the sun. The shoots must be tied up several times, to keep them on their own poles.

Diseases.—The worst bean enemies are rust and blight. In new soil, with good weather, these troubles seldom appear. During prolonged wet weather there seems to be no help for them. Spraying with Bordeaux mixture is a preventive. The spraying should be done in advance of blossoming. The seed is sometimes soaked in Bordeaux mixture for an hour where rust is anticipated. Prevention is better than cure, and new soil and fresh seed are the best precautions. Diseased vines should be burned.

Insects.—The weevil which attacks the bean is closely allied to the pea weevil. Some practical people say there is no remedy known; others recommend heating the beans to 145° for an hour; others use bisulphide of carbon in a closed vessel, along with the beans.

Profits.—By far the largest cash receipts per acre are obtained by selling beans in their fresh state; preferably in the pods. The production of bush beans (pods) may run up to 75 or 80 bushels per acre, or even more. Lima beans are more profitably sold in the pods than shelled, though some markets demand the shelled article. The consumer gets a fresher and better article in the pods, and the producer is saved much trouble, and this method should be encouraged. Beans should be cooled, if possible, before shipment in bulk to distant markets, thus avoiding danger from heating, moulding and spotting.

BEETS.

Beets are produced in enormous quantities by market gardeners near all large cities, both under glass and in the open ground. They also have a place in the farm garden, as they are of easy culture.

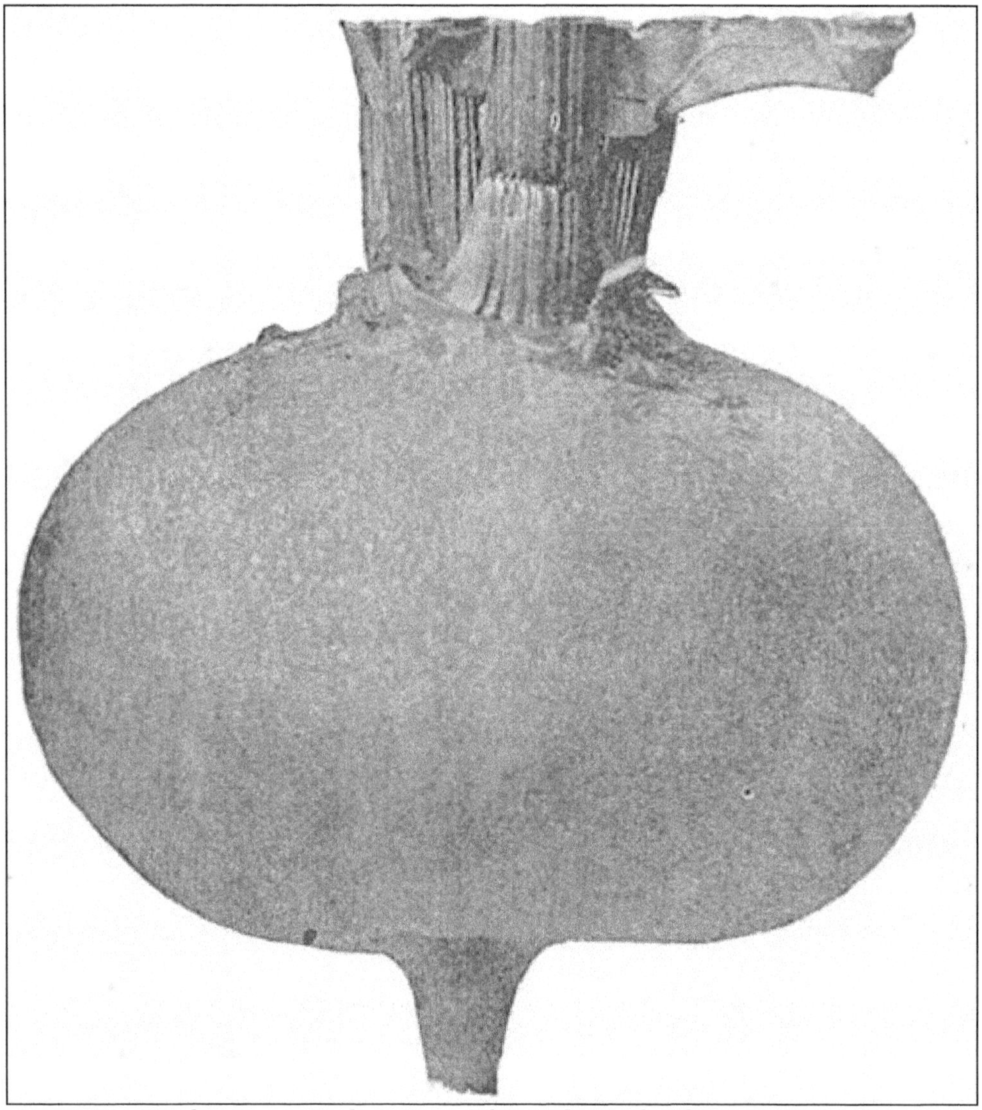

Crosby's Improved Egyptian, the Earliest Blood Turnip Beet.

Excellence in the table beet depends partly on variety, but mainly on the quickness of growth. Sweetness and succulence result from high culture in rich, mellow soil.

Mangels and sugar beets, of course, have a place on every farm, for stock-feeding purposes, and table beets may also be grown, if good soil is available, for market purposes. The winter-keeping sorts are frequently in demand, and may be included among the farmer's cash crops.

No amount of stable manure is excessive in beet-growing. Partially rotted manure is best. For horse culture the rows should be 3 feet apart. Five to six pounds of seed will plant an acre.

Planting.—Planting may be done as soon as the ground can be worked in the spring, as the beet is hardy, and not injured by a little frost; and successional plantings may be made until June. The June sowing will produce autumn beets, which can be stored for winter use or sale.

It is well to soak the seed in tepid water before planting; it should be scattered thinly in the rows and lightly covered. In dry weather the soil must be pressed firmly on the seed, to insure sufficient moisture for germination. The plants in the rows should be thinned out to 3 or 4 inches.

It is very important to remember that the more space each plant has about it the sooner will it reach a marketable size. Beet plants standing 5 inches apart in the row will be ready long before plants standing only 2 inches apart.

Beets vary in shape very considerably. Some are round and some are long, with intermediate grades. The turnip-shaped beets are the earliest, while the half-longs and longs are the heaviest. For market purposes, if sold in bunches, the round ones are the most profitable.

The color of the foliage varies greatly; but the color of the leaf is not always typical of the root. Some of the blood beets have green leaves. There are many shades and colors of the roots, from deepest blood red to white, with zones of pink. The beet is an excellent and highly esteemed article of food, and is always in demand.

Beets.—For earliest, we especially recommend Crosby's Improved Egyptian and Surprise; for winter, Ford's Perfected Half-Long. Please see "Johnson & Stokes' Garden and Farm Manual."

Marketing and Storing.—A bunch contains five, six or seven beets, with tops tied together and superfluous leaves cut off. The bunching and topping may be done in the field, and the bunches afterward washed in a tub of water, by means of a scrubbing brush. It always pays to send roots to market in a clean and attractive condition.

Winter storage in cellars, under sand, is often practiced; or the beets may be kept in pits in the open ground, covered with straw and earth.

Enemies.—The beet is remarkably free from enemies of any kind. The root sometimes cracks, and is occasionally attacked by insects, but the farmer or gardener has little to fear if soil be good and weather be favorable. All farmers attending market should have a few beets to help make up the weekly load for the wagon.

CABBAGE.

Early cabbage is not a farm gardener's crop at the North, though in the Southern States the early varieties can be grown by farmers for shipment to the great Northern markets. The Northern farmer, unless provided with glass, usually finds more profit in the later and larger sorts, which mature in autumn.

Soil.—Rich, loamy soil, containing much clay, is best for this vegetable, which is a rank feeder. Large amounts of manure are demanded. The manure is best applied in a partially rotted form, as fresh manure of any kind (especially hog manure) is liable to produce the disease or deformity known as club-root, the spores of the disease apparently being in the fresh manure; though land too long cropped with cabbage is likely to produce the same disease without the application of fresh manure of any kind.

Early Jersey Wakefield Cabbage.

Seed.—It is of especial importance that good seed be planted, as cabbage varies so much and shows such a disposition to go back to undesirable types that great dissatisfaction and loss attend all experiments with poorly-selected seed. The choice of seed not infrequently determines the size and success of the crop. Expert cabbage growers are well aware of this fact.

Johnson & Stokes 'Market Gardeners' No. 2, Valuable for Early Summer or Winter Cabbage.

Planting.—The manure should be broadcasted, and an ample amount used, with a high-grade fertilizer in the row. The young plants, previously started in a seed-bed, should (at the North) be set out in July. The seed for late cabbage is planted in May. A quarter pound of seed will give enough plants for an acre.

The rows should be 4 feet apart, and the plants 2½ feet apart in the rows. These distances favor good cultivation and quick growth. In some parts of New England the seed is sown in the open field, in rows where the cabbage is to grow, but the practice of transplanting from seedbeds is found most satisfactory.

The rainfall here usually insures a fair crop of cabbage, but any crop which requires transplanting in midsummer is liable to delay or injury in case of protracted dry weather. Hence, irrigation is desirable. At the distances just recommended for planting (4 × 2½ feet) there would be 4,356 plants to the acre. In the case of such varieties as Johnson & Stokes' Earliest and Jersey Wakefield cabbage, where the number of plants per acre would be perhaps 10,000, the Michigan Experiment Station obtained 5,000 more marketable heads per acre under irrigation than where water was not used upon the growing crop. (This fact is mentioned in a book on irrigation just issued by the publishers of this book).

Varieties.—The earliest varieties of cabbage have small, conical heads; the midsummer sorts mostly round heads; and the late or drumhead sorts have large, flat heads. There are cabbages which never head; as, for instance, the collards of the South; and there are varieties of crinkled-leaf cabbages, known as the Savoy types. The kales are closely related to the cabbages. Both cabbages and kales have purple-colored forms, sometimes called red forms.

Cabbage.—For early varieties for the South, we recommend Johnson & Stokes' Earliest, Early Jersey Wakefield and Charleston Wakefield; for both early and late in the North, Johnson & Stokes' Market Gardeners' No. 2, Louderback's All the Year Round; for late varieties for the North, New Rock Head Winter, Johnson & Stokes' Matchless Flat Dutch, Danish Ball Head. The Johnson & Stokes' Hard Heading Savoy Cabbage is of rare excellence. For descriptions of the many varieties of cabbage, please see "Johnson & Stokes' Garden and Farm Manual."

Cultivation.—Thorough horse cultivation between the rows should be supplemented by a hand-hoe between the plants in the rows. The cultivation must be good and continuous until the heads begin to form.

Diseases and Insects Enemies.—Club root has been mentioned. It is a fungous trouble. The best remedy is new ground. The black flea on very young plants can be conquered with air-slacked lime or wood ashes. The cut worm is troublesome only in spring; not with late cabbage. The root maggot is sometimes very destructive, both with cabbage and cauliflower. New ground is the most satisfactory remedy. Green aphides or lice often follow lack of strength in the cabbage. Pyrethrum powder, air-slacked lime, kerosene emulsion, etc., are used as remedies for lice. The pyrethrum may be used dry or in water, at the rate of a tablespoonful to two gallons. The green cabbage worm, one of the worst of all enemies, can be pretty effectually checked by means of air-slaked lime dusted over the leaves. Other caterpillars yield to the same treatment.

Bursting.—Bursting of cabbage heads is caused by a second growth, the result, perhaps, of continued wet weather, or warm weather following cold weather. The best remedy is to cut part of the feeding roots, either by close cultivation or with a hoe.

Selling.—Cabbage prices vary between extremes that are far separated. Early cabbage usually sells at a good profit. Summer and autumn prices may be low. Winter and spring prices are almost always fair, and occasionally extra. Pennsylvania farmers sometimes ship to wholesalers in the cities and sometimes sell at public sale in the open field, in the autumn, just as the crop stands. The latter plan is an excellent one, where auction prices warrant it. It avoids the cost and risk of storage, as each buyer removes and stores his purchase.

Storage.—Cabbage will bear much freezing without injury. The art of winter storage is to put it where it will have the fewest changes of temperature, and where it will be cool and damp without being wet.

The most common practice is to cover two or three rows of inverted heads, with roots attached, with from 6 to 12 inches of soil, making provision for good drainage by ditches on both sides of the wedge-shaped heap.

Cutting Johnson & Stokes' Earliest Cabbage for Market. Photographed June 1st, in the Field of Messrs. Myers & Bowman, the well-known Philadelphia Market Gardeners. This was the first Home-grown Cabbage in Philadelphia Markets.

This system may be modified so as to include six or more rows of inverted cabbage, the heap being flat instead of wedge-shaped on top. It does not turn water so well, but in practice is usually satisfactory. A good plan is to use about 6 inches of soil, and to add straw or litter as the cold increases.

Under a steady low temperature it is no trouble to keep cabbage through the winter, but it is hard to provide against the many changes of our variable climate.

Johnson & Stokes' Matchless Late Flat Dutch Cabbage.

Where heads are to be carried over for seed, or where it is intended to head up soft cabbages during the winter (a feasible thing) the roots are set downward instead of upward. If care be taken to remove the roots without much injury, they may be set in furrows or trenches, and the earth heaped over the cabbages just as in the several ways above mentioned, and they will make decided growth during their life under ground. In fact, a cabbage with any sort of immature head in November will, under proper management, be in good marketable condition in March or April.

Solid freezing in the trenches is not necessarily destructive, but if the temperature falls much below 15° (at the point occupied by the heads), there is danger that they will perish. They may be in good edible condition after such severe freezing, but the chances are that they will fail to grow if set out for seed. The cabbage decays with a strong, offensive smell when its tissues finally break down after repeated changes of temperature and moisture. A uniform temperature is favored by the use of earth in storage, and though storage in buildings and cellars is quite feasible, there is nothing better or cheaper than the soil of the open field.

If the crop is not all to be marketed at one time, it is well to make a number of separate trenches, so that each can be wholly cleared of its contents at a single opening. These trenches and ridges must be made upon dry ground, where there is no standing water.

For Stock.—Cabbages make good food for cows, but should be fed after milking; and frozen cabbages should never be fed in any considerable quantity, as they are liable to cause hoven or bloat.

CARROTS.

A sandy soil or light loam is best for carrots, but they will grow anywhere under good culture. Enormous quantities are grown by the market gardeners, both under glass and in the open ground, for use in soups and for seasoning purposes. The short or half-long varieties are demanded by this trade.

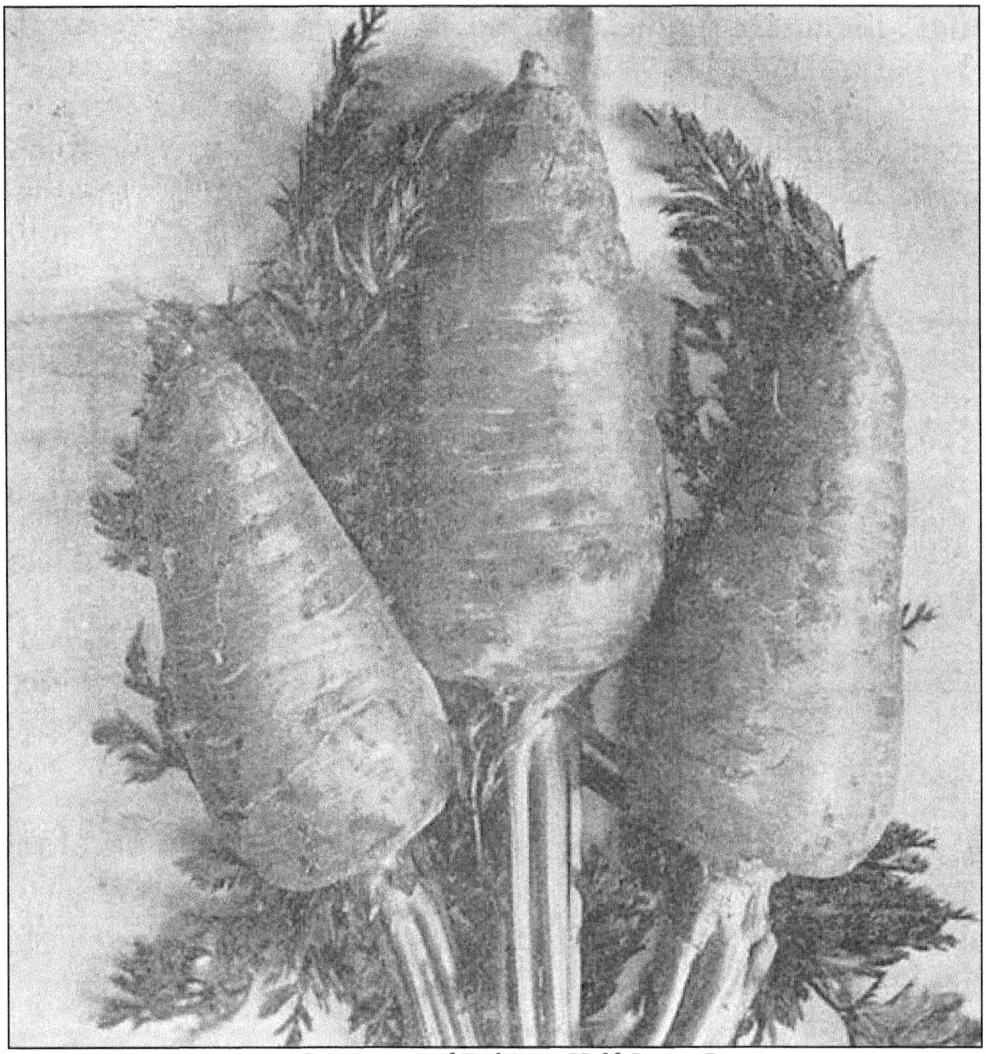
Average Specimens of Rubicon Half-Long Carrots.

Farm gardeners will do best with half-long and long kinds, unless a special demand calls for the smaller carrots. The large half-long and long ones are suited to both culinary and stock-feeding purposes.

It requires from three to four pounds of seed to the acre, depending on the distance between the rows. The plants should be from 3 to 5 inches apart in the rows, and the rows as near together as is feasible for horse work. Clean culture is demanded. The seed must be planted shallow, and may go into the ground as early as it can be worked in the spring, and from that time until the middle of June. The only danger about late planting is the possibility of dry weather.

The carrot is quite free from insect or other enemies, as a rule, and its culture is not difficult. It demands thinning and hoeing after the plants are well above ground, but no extra attention of any kind.

The winter storage is the same as for beets or turnips, either to be put away in earth-covered heaps or preserved in a cool, non-freezing root cellar.

The so-called Belgian carrots (both yellow and white) are used only as stock food; though the other sorts, such as Rubicon, Danvers and Long Orange, if in excess of market demands, are equally good for stock. Cows and horses are fond of them, and they are most wholesome. The farm gardener should raise them, however, for their cash value in the produce markets. The carrot is in high favor with good cooks everywhere.

The carrot does not demand excessively rich ground; in fact, too much manure tends to stimulate the growth of the top at the expense of the root, and fresh manure makes the root rough.

The smaller carrots are bunched and sold like radishes or early beets. The larger kinds are sold by measure—about 60 cents or more per basket at this time (January, 1898). This is at the rate of $1.50 per barrel, or about $300 per acre. The crop is a good one, if near a market where carrots are demanded.

CARROT.—We especially recommend Rubicon Half-Long for market or stock. See "Johnson & Stokes' Garden and Farm Manual."

SWEET CORN.

There is no money crop more available to the farm gardener than sweet corn. It will grow anywhere, and the young ears are always in demand. Any sod land plowed shallow will yield a crop of sweet corn. It is easy in this latitude to have an unbroken succession of marketable ears from July 1st to October 1st, or even somewhat earlier and later.

Shallow plowing and the use of a little fertilizer or compost in the hills will put the ground in order. A complete fertilizer is best. A compost containing hen manure is excellent.

Planting.—Eight or ten quarts of seed are required to plant an acre of corn in hills, allowing for replanting of what is injured by grubs or other causes. The larger varieties should be planted 4 feet by 3; the rows 4 feet apart and the hills 3 feet apart, with not more than three stalks in a hill. The smaller varieties may be grown much closer—3 feet by 2. Any method may be used in laying out a corn field that will give each stalk (of the large kinds) the equivalent of 4 square feet of ground space. The dwarf sweet corns demand about half that space.

Varieties.—The sweet corns require from sixty to eighty days to produce ears fit for boiling. The earliest varieties are small, and are lacking in sweetness as compared to the best intermediate types. Still, the early prices are so much better than midsummer prices, that the early varieties will always be grown for market. Indeed, the best profits of the business are from the extra early and extra late sales.

Sweet corn should not be grown by shippers who are distant more than twenty-four hours from market, as the ears lose quality and flavor soon after being pulled from the stalks. Forty-eight hours from market is an extreme distance, but is feasible if the ears can be chilled in a cold storage house previous to shipment; otherwise, they will heat and spoil. Even when designed for a near-by market a load of sweet corn ears may heat and spoil during a single night. It is best to scatter them upon the grass, if pulled during the afternoon for shipment the following morning.

The most profit to the grower will be found in ears which are not too large, as corn is often sold by the dozen, the large sorts being too weighty.

The early kinds, though small, can be planted closely, and a large number of ears secured; and they are out of the way so soon that the ground can be used for celery or other late crop. Celery can be set out between the rows of corn, and thus be shaded to some extent during the critical period following transplanting.

The Evergreens, Early and Late, and the shoe-peg types, such as Country Gentleman and Zig Zag Evergreen, are among the sweetest of all. The grains are of irregular shape and arrangement, and the appearance of the ears is not altogether prepossessing. When once known, however, they are in demand by consumers.

The red-cob corns should be cooked by dropping into boiling water. If cooked slowly, the red color of the cob affects the appearance of the grains.

SWEET CORN.—For first early, we recommend Burlington Hybrid and Mammoth White Cory. The former closely resembles a true sugar corn in appearance. For second early, Early Champion and New Early Evergreen; for late, original Stowell's Evergreen, Country Gentleman, Zig Zag Evergreen. See "Johnson & Stokes' Garden and Farm Manual" for descriptions of varieties.

Cultivation, Enemies, etc.—Shallow culture, frequently repeated, is demanded by sweet corn. The growth at first is timid and slow; afterward, if well cultivated, the stalks grow with great rapidity and vigor.

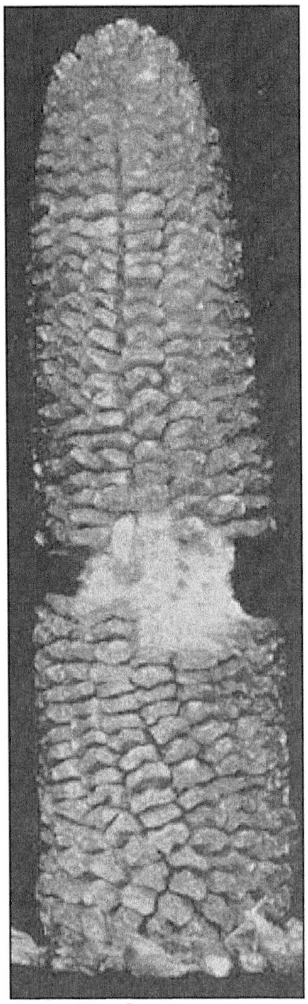

New Early Evergreen Sweet Corn.

To make the most of the stalks, they should be cut as soon as possible after the last ears have gone to market and fed to stock. Sweet corn stalks when dry make excellent fodder.

The main enemies of corn are the cut worm, which is only troublesome in spring; a fungus which attacks the ears and which is always most prevalent on the small, early sorts; and a worm which cuts and injures the grain while the corn is in milk. Crows sometimes pull up the seeds, but can be disposed of by scattering a little yellow corn on the surface of the ground around the edges of the field. As the crow destroys many cut worms, it is better to feed him with corn than to shoot him.

The prevalence of fungus-troubled or smutty corn is probably a symptom of weakness, the result of planting too early, or of too much wet weather. All plants that are weak are liable to fungus attacks, and it is the early corn that suffers most. This corn is often planted before the ground is sufficiently warm, and there is a consequent weakness of growth. Indian corn, at Philadelphia, should not be planted before May 10th, and yet it is not uncommon to see gardeners planting sweet corn two weeks earlier. They say they are "going to risk it." The result may be a good crop of corn, or it may be a crop of worms and fungus. Of course, the high price of the first corn in market is the excuse for the unseasonable date of planting. But it is not quite fair to blame the seed or the variety of corn for what is partly the result of the gardener's impatience. All traces of smut on corn stalks should be burned, and not allowed to be fed to cattle.

The Corn Worm.—Far more destructive and disastrous than smut is the corn worm (Heliothis armiger). This is the cotton worm of the South, there called boll worm. It is also sometimes called the tomato worm. It is the larva of a day-flying moth.

The difficulty in dealing with it is that when in the corn ear it is out of the reach of poisonous applications of any kind. Its depredations are extensive, especially in early corn. It prefers corn to all other foods, and cotton planters protect their crops by planting early corn in the cotton fields and then destroying the corn and the worms within the ears.

The best remedy at the North is to feed all wormy ears to pigs; and to plow the corn land in autumn, when the insects are in the pupa or chrysalis state. If turned up by the plow, it is believed that they mostly perish. The worms are said to be cannibals, eating each other to a great extent.

This worm is, perhaps, the greatest enemy with which the grower of sweet corn has to contend. The plan of feeding wormy ears to pigs offers the double advantage of destroying the enclosed pests, while at the same time fattening the pigs.

Successional Planting.—The skillful farmer will arrange successional plantings of corn, beginning (at Philadelphia) May 10th and ending about July 10th. The first and last plantings should be of the early sorts; the intermediate plantings of the full-sized varieties.

Profits.—Profits depend on location. The size of the crop should approximate 1,000 dozens of ears per acre, and the gross receipts should be $100 to $200 per acre, more or less, above the value of the fodder.

Suckering.—Time is often spent in pulling the suckers from the stalks of sweet corn. Such time is wasted. If the suckers are let alone they will not reduce the number or quality of the marketable ears.

HORSERADISH.

Farmers who have soil that is both rich and deep can find profit in growing horseradish on a large scale, in connection with early peas, beans or sweet corn. The sets are planted in May, in the rows between crops, and after the crops are removed the horseradish makes its main growth. It is perfectly hardy, and comes on rapidly during the late summer and autumn months. Where the ground is not strong enough to produce large roots the first year, the business will not prove very remunerative.

The Sets.—Horseradish sets are made by cutting small roots (¼ to ½ inch in thickness) into pieces 6 or 8 inches long. The upper end is cut square off; the lower end with a slope. This is to get them right end up at planting time. The small roots are available in quantities in the autumn, when the large roots are trimmed for market. The sets are kept in sand during the winter, or buried in the open ground, in a carefully-marked spot, where they can be easily found in the spring.

If planted 2 feet apart in rows 3 feet apart, each plant will represent 6 square feet of space, and, hence, about 7,300 sets will be needed for an acre.

The method of planting is to strike out rows, and with a long dibber or crowbar make holes 8 or 10 inches deep. A set is dropped into each hole and the earth pressed about it.

Shoots will soon appear above the surface, and when the early crop has been removed from the land, the horseradish should be well cultivated once or twice. Little further attention is needed.

The roots should be lifted the same year, in December, and stored in an earth-covered heap or pit, or else in sand in a root cellar. The small lateral roots should be saved for the next year's sets. There is a good demand for horseradish, both wholesale and retail; but prices should be ascertained before going into the business in a large way.

Good roots, after trimming and washing, should weigh half a pound or more each.

PARSNIP.

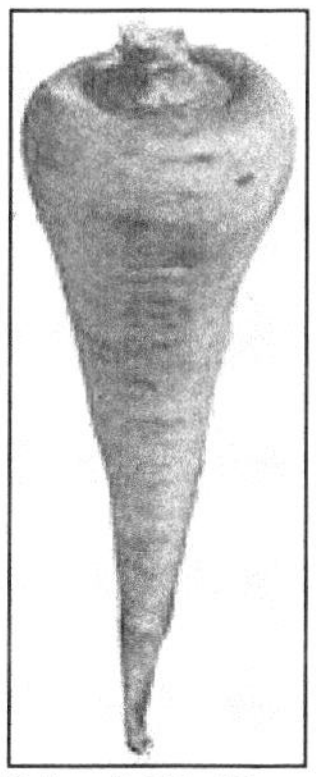

Ideal Hollow Crown Parsnip.

The cultural requirements of the parsnip are quite similar to those of the carrot. Any soil that is deep mellow and moderately rich may be used for parsnips. Fresh manure is to be avoided, as it makes the roots rough.

The seed should be planted in early spring, while the ground is moist, as it germinates very slowly. It should be covered to a depth of half an inch, and the soil pressed down firmly. The plants must be thinned out to stand 3 or 4 inches apart.

The parsnip is a vegetable of a perfectly hardy character. It may remain in the ground, just where it grows, all winter. The flavor is said to be improved by hard freezing, and no amount of freezing will hurt it.

It has a high value as human food, and is demanded in large quantities in some markets. It also has a high value as a stock food, especially for cows. It should be fed after milking, in quantities not sufficient to taint the milk. The price is variable, but about the same as the carrot.

PARSNIP.—We recommend Ideal Hollow Crown. For description, see "Johnson & Stokes' Garden and Farm Manual."

SALSIFY OR OYSTER PLANT.

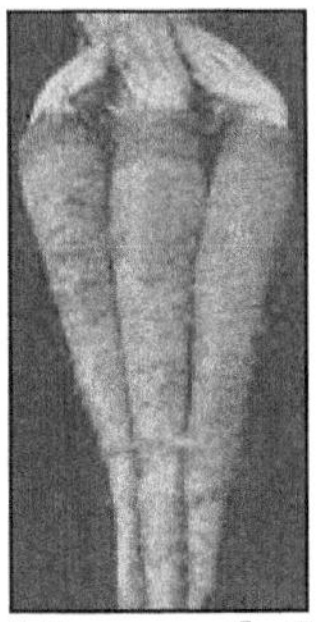

Mammoth Sandwich Island Salsify as Bunched for Market.

Salsify, oyster plant or vegetable oyster is a root of easy culture and of high food value. In shape it resembles the carrot and parsnip, and is as perfectly hardy as the latter. The seed should be sown an inch under the surface, in spring, in rows 2½ to 3½ feet apart, and the plants thinned to stand 5 inches apart in the rows. The culture is the same as for parsnips. Fresh manure must be avoided, as it makes the roots ill-shaped. The roots, under good treatment, will exceed an inch in diameter, and may attain a size of 2 inches or more. They may remain in the ground over winter, to be taken up whenever the frost permits or they may be taken up in late autumn and preserved in sand in a cellar. Good salsify is in demand where its merits are known.

THE POTATO.

The cultivation of the potato is so well understood by every American farmer and gardener that it seems unnecessary to discuss the details of cutting the tubers, planting, cultivating, harvesting, etc. The weak points of potato culture are most commonly the fertilizing and the treatment of diseases. These will be briefly discussed. As to lack of moisture, to be remedied by artificial watering, the reader is referred to our new book, entitled, "Irrigation by Cheap Modern Methods," in which a case is mentioned where water alone made a difference of 129 bushels per acre in the crop.

POTATOES.—Best for the South, Bliss Triumph, Pride of the South, Crown Jewel, Early Thoroughbred. General crop in the North—Houlton Early Rose, Table King, Late Puritan, Rural New-Yorker No. 2. For descriptions of these and other varieties, see "Johnson & Stokes' Garden and Farm Manual."

Fertilizing.—A ton of potatoes (33-1/3 bushels) contains 4·2 pounds of nitrogen (equal to 5·1 pounds of ammonia), 1·5 pounds of phosphoric acid and 10 pounds of potash. This shows that nitrogen and potash are the elements mainly abstracted from the soil by a crop of potatoes. An analysis is not an infallible index of what must be applied to any soil, for that soil may be naturally rich in some one fertilizing element and deficient in the others. Only experiment will
 determine what is best. But a knowledge of the analysis of the crop is necessary to intelligent experimentation. Nitrogen and potash will evidently be demanded in most cases, yet the Ohio Station recently reports that "phosphoric acid has been the controlling factor in the increase of the potato yields" in the trials made there. This shows how greatly soils vary in their requirements.

Harvesting Seed Potatoes near Houlton, Aroostook County, Maine.

Barnyard manure would answer all purposes and would be an ideal potato fertilizer, except for the fact that it so often carries with it the spores of such diseases as blight, scab and rot. Still, barnyard manure in a partly rotted condition is very widely used by potato growers.

Clover sod is an excellent source of nitrogen, as heretofore explained. The clover is, perhaps, the best of the leguminous crops for green manuring purposes. Many successful potato farmers depend largely upon clover, supplementing it with a small amount of high-grade complete fertilizer in the rows.

Where phosphoric acid is necessary, it can be had in the form of ground bone or acidulated rock, and potash can be had in the form of sulphate or as kainit. Where the scab is prevalent, it may prove better to use kainit, on account of the salt which it contains, as will be presently explained.

Planting.—It requires from seven to ten bushels of tubers to plant an acre. Some growers use as much as fifteen bushels. The date of planting, depth, distance between rows, etc., are details for individual determination. Flat culture is better than ridge culture, so far as conservation of moisture is concerned. It is important that good Northern-grown seed be planted; tubers which have not lost their strength by excessive sprouting. Storage in a cool, dark, dry place is best for potatoes. Whether planted early or late, or at successional dates, must be determined by the market requirements of the grower.

Varieties.—The varieties of potatoes are many, and while it is wise to experiment in a small way on new kinds, it is best to depend for business purposes on standard sorts that have been fully tested.

Irrigation.—After the farmer has exhausted his best efforts in the preparation and fertilization of the soil, and after good seed has been planted and the best possible culture given, there may come a season of prolonged drouth that will defeat his purpose of securing a large crop. This result is not common, but neither is it rare; and where farmers are looking toward the high culture of certain special crops, it would be well for them to consider the matter of artificially watering their potato fields.

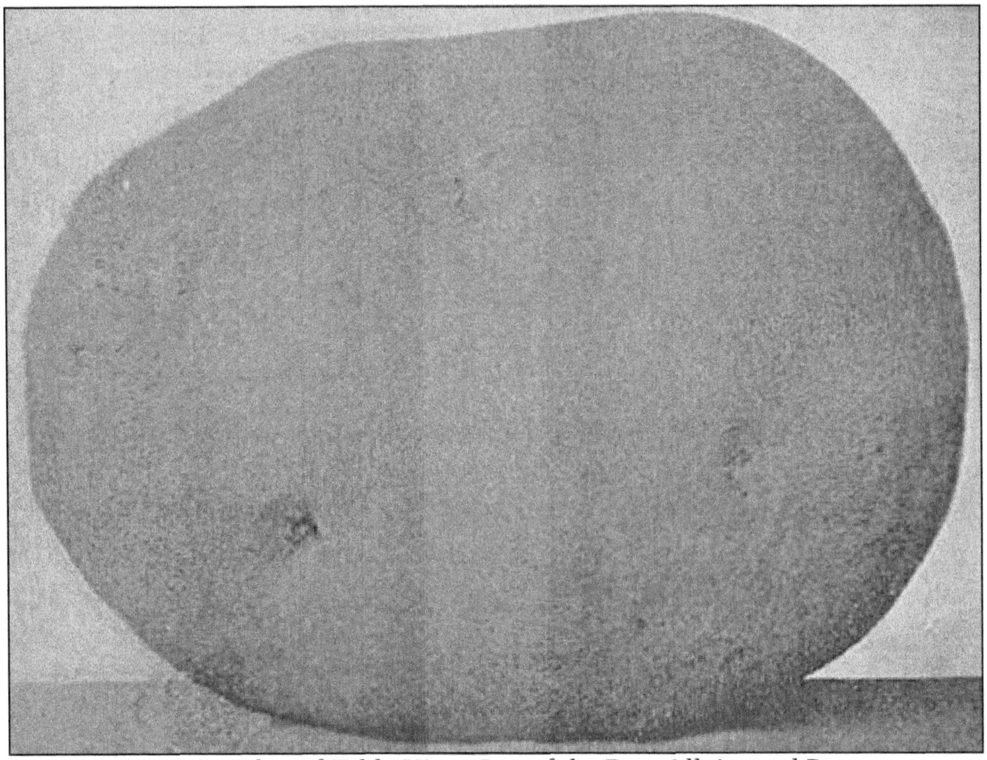
An Average Tuber of Table King, One of the Best All Around Potatoes.

Diseases and Enemies.—Not counting dry weather, which sometimes robs the farmer of two-thirds of his crop, there are four diseases which exert a disastrous influence on the potato, and which are liable to occur any season. Two of these diseases are of the leaf and stem and two of the tubers.

The two leaf troubles are respectively known as blight or downy mildew and the Macrosporium disease. The two tuber troubles are scab and rot.

Leaf Blights.—No attempt will be made here to separately describe the two leaf diseases. Both destroy the foliage and check the further growth of both vine and tuber. The leaves turn brown or black, and the stem quickly wilts and falls. There can be no growth of tuber without vigorous health of vine. Spraying with Bordeaux mixture, in advance of the occurrence of any disease, is recommended.

Bordeaux mixture for this purpose is made by using six pounds of copper sulphate and four pounds of quick lime, dissolved in separate wooden vessels, and the lime water poured into the dissolved blue stone. This should be diluted with water sufficient to fill a forty-five gallon barrel. Paris green to the amount of from one-quarter to three-quarters of a pound to the barrel should be added, to destroy beetles and other insects.

The vines should be sprayed five or six times, beginning when they are 6 inches high, at intervals of ten days or two weeks. During rainy weather the spraying should be more frequent than during clear weather. The object is to prolong the life and vigor of the vines. The cost of the five or six sprayings, including labor at $1.50 per day, is put at not above $6 per acre, while the crop at stake may be affected to the extent of scores of bushels.

Scab and Rot.—The evidence about scab and rot is still contradictory, but it is likely that these diseases will presently be under control.

At the New Jersey Station, Professor Halsted completely conquered scab with an application of 300 pounds of flowers of sulphur per acre scattered in the rows, while the same treatment at the Ohio Station was less successful. At the latter station benefit was found in the use of salt, kainit, sulphate of potash, etc.

The various experiments and observations on potato scab and rot seem to indicate that scab flourishes best on a soil inclined to be alkaline, while rot is most prevalent on a soil inclined to be acid. The use of lime increases scab, while the use of kainit diminishes it.

The best practice, therefore, under present knowledge, would be to use clean seed on new ground, avoiding fresh stable manure. Clean seed can be had by treating tubers with corrosive sublimate. This substance is dissolved to the amount of 2¼ ounces, in two gallons of hot water, and (after standing a day) diluted with water so as to make fifteen gallons. In this solution the uncut seed potatoes should be soaked for an hour and a half. All unplanted seed potatoes should be destroyed, as the corrosive sublimate is highly poisonous.

The use of sulphur, as recommended by Professor Halsted, will prove entirely satisfactory in some soils. In others, the use of kainit or sulphate of potash or acid phosphate would no doubt be found preferable.

Where soil is badly affected with disease germs, it is unquestionably better to seek a new field than to attempt to disinfect the old one. A rotation of crops will probably restore diseased land to health more cheaply and more thoroughly than any other process.

Profits.—Of potato profits it is not necessary to speak, except to remark that it costs but little more to produce 300 bushels to the acre than 100 bushels. There can be no doubt whatever that it pays handsomely to spray potato vines with the Bordeaux mixture.

PUMPKINS AND SQUASHES.

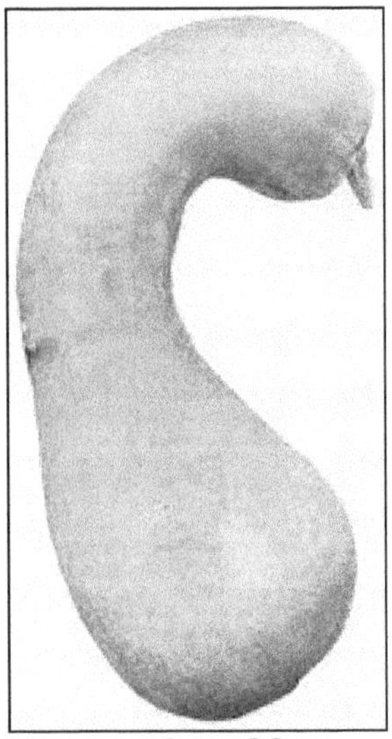

Mammoth Golden Cashaw Pumpkin, One of the Best for Market or Stock Feeding.

There is no clear dividing line between pumpkins and squashes, as they belong to the same botanic family—the Cucurbita. Some members of the group are clearly pumpkins, and others just as clearly squashes, but when an attempt is made to draw a sharp line between them, we get into difficulty. In general terms the pumpkin has a soft rind or shell and the squash a hard rind. But even this thumbnail test is not infallible.

These vegetables belong on the farm, on account of the large ground space occupied by the vines. Pumpkins may be economically grown in corn fields, the seeds being planted along with the corn—one pumpkin seed to every fourth hill. No special care is needed besides the cultivation given the corn.

Farmers should give far more attention to growing squashes, as they are much superior to pumpkins in food quality, both for the table and for stock.

There are numbers of excellent squashes now catalogued by the seedsmen which many farmers have never tried, but which are worthy of cultivation for market purposes. When a farmer by experiment has found a high-quality squash adapted to his soil, he has put himself in possession of a product of permanent market value.

PUMPKIN.—We especially recommend Mammoth Golden Cashaw and Winter Luxury. For descriptions, see "Johnson & Stokes' Garden and Farm Manual."

SQUASH.—Early varieties—Mammoth White Bush Scalloped, Giant Summer Crookneck. Winter-keeping varieties—Sweet Nut, Faxon, Chicago Warted, Hubbard, Early Prolific Orange Marrow. For descriptions, see "Johnson & Stokes' Garden and Farm Manual."

TOMATOES.

Tomatoes may justly be rated among the leading crops available to farm gardeners. There is always a brisk market for selected, carefully-washed tomatoes, packed in new baskets. Such produce is seldom offered in excessive quantities.

Any good corn land will produce good tomatoes. Excessive manuring is likely to stimulate the vines at the expense of the fruit. A little complete fertilizer or compost in the hills is desirable.

Tomato seed of early varieties should be started under glass. The seed is sown on heat and the plants once or twice transplanted, and put in the open ground as soon as danger of frost is over. Little is gained by setting out too early, when the ground is cool. The tomato is of tropical origin, and makes rapid growth only at a temperature of 65° or upward. Indeed, it is suspected that one of the worst diseases to which the tomato is liable, the blight, is encouraged, if not wholly caused by too early planting in the open ground.

Varieties and Planting.—At 4 feet apart each way, it will require about 2,700 tomato plants for an acre of land. In open field culture the tomato is always allowed to lie upon the ground. In garden culture, it is often tied to stakes or supported on trellises. Three ounces of seed will raise sufficient plants for an acre.

There are many varieties of tomatoes, including the early and late market sorts, the yellow kinds, and the little pear-shaped and plum-shaped tomatoes, both red and yellow, used in pickling. The ideal market tomato is one of medium size and smooth shape. It must have firmness and depth, and the quality of ripening evenly all over. There should be neither greenness nor wrinkles around the calyx, nor should the fruit be of irregular shape. As to color, it is a matter of taste and neighborhood preference. Some markets demand red and some purple fruit.

TOMATO.—We recommend, for earliest, Atlantic Prize and Money Maker; for second early and main crop, Brinton's Best, New Fortune; for late, Brandywine, Cumberland Red, Stone. For descriptions, see "Johnson & Stokes' Garden and Farm Manual."

Successional Planting.—If the first tomato plants be set in the open ground (at Philadelphia) May 15th to 20th, there should be at least one and preferably two later crops, because young, vigorous plants yield the most and best fruit. It is good practice to sow tomato seed in the open ground, say about middle of May, and again somewhat later. These out-of-door plants will come forward very rapidly, and will be ready to produce late summer and autumn crops.

Atlantic Prize Tomatoes, as they Appear for Sale on Fruit Stands, etc., during the Spring Months.

Cultivation.—The tomato is of the easiest cultivation, and will grow even under neglect, but it so abundantly repays attention, that no farmer can afford to be careless about the matter. The nights of May are cool in the North, and the newly-set plant at first makes little growth. Cultivator and hand-hoe should both be kept in motion during this period, and in June also. In the latter month the tomato will make a sudden leap toward maturity, and will yield ripe fruit in July. The out-of-doors cropping season lasts for three full months. The tomato is now grown under glass almost everywhere, and is to be had in the market during all the months of the year.

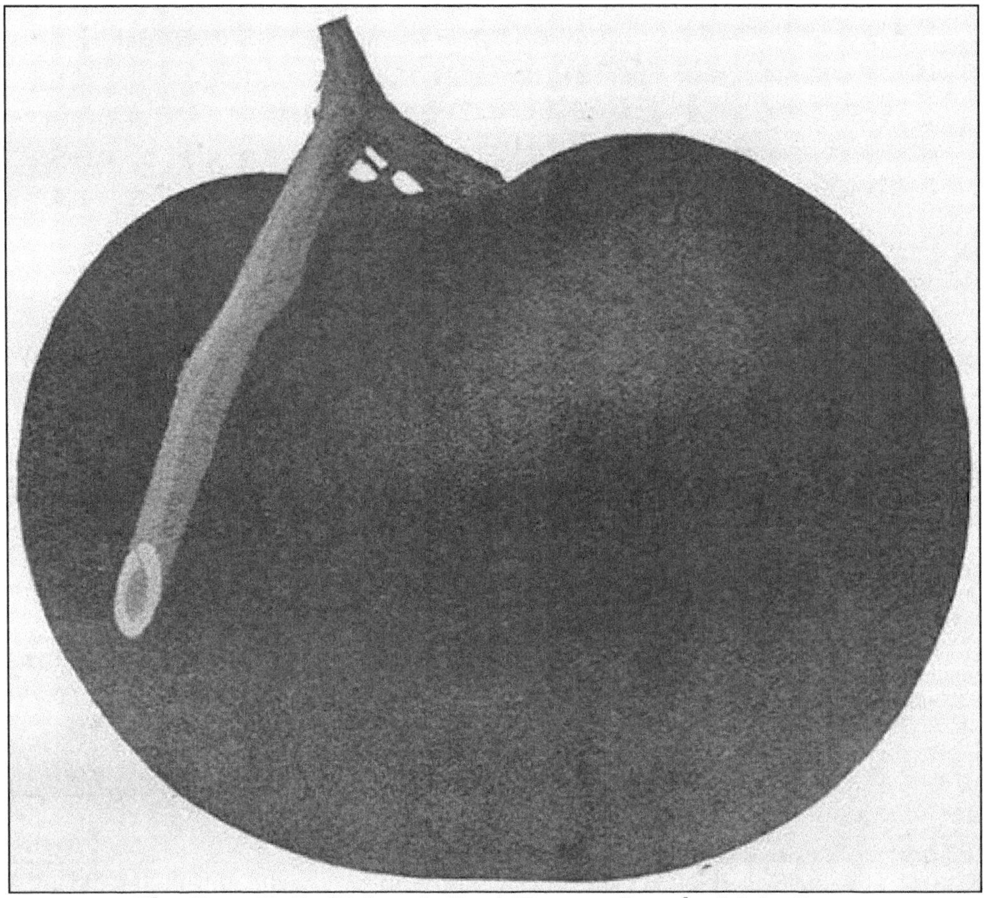
The Great B. B. (Brinton's Best) Tomato, Best for Main Crop.

The out-of-doors season is profitably prolonged by picking all the mature or nearly mature fruit when the first frost comes, in October, and placing these unripe tomatoes on straw in a cold frame. Covered with straw and with the sash to keep out frost, the fruit ripens in a satisfactory manner for several weeks. Such a frame must be well ventilated or the tomatoes will rot rather than ripen.

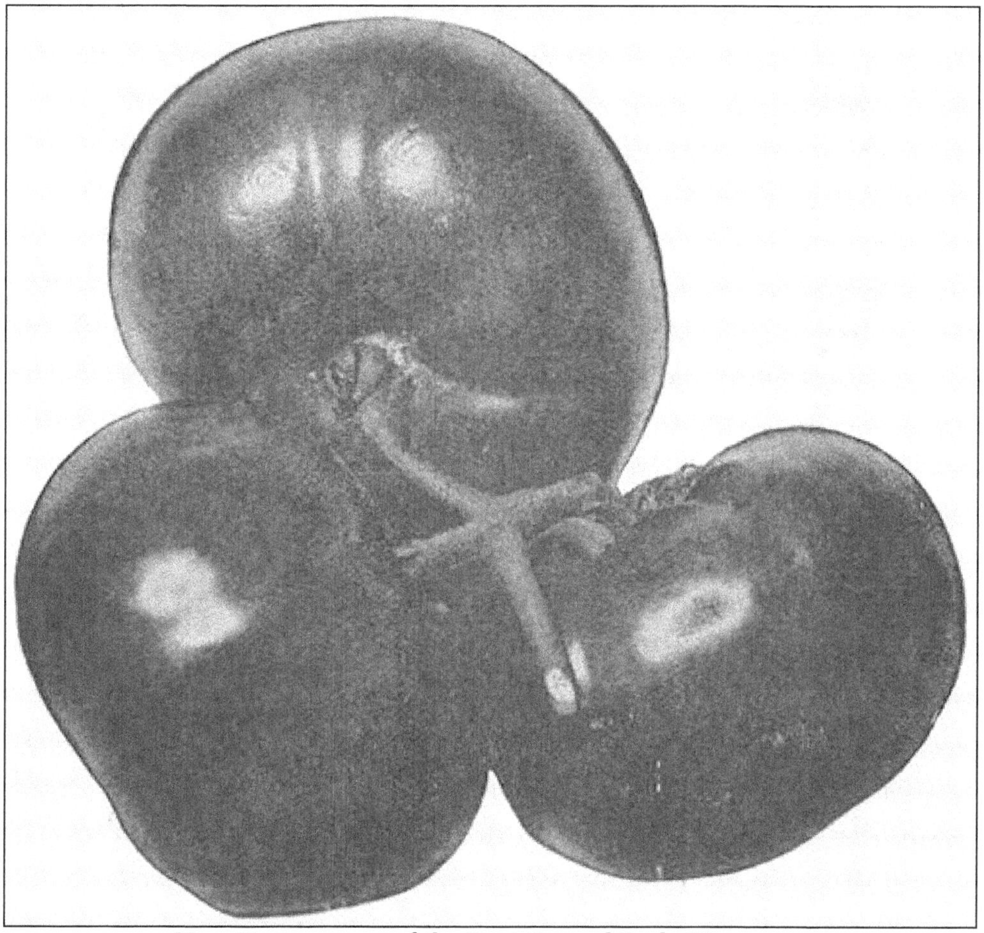

New Fortune, one of the Best Second Early Tomatoes.

Diseases and Enemies.—Tomato diseases, fortunately, are not numerous. Blight sometimes sweeps off a whole field of early-set tomato plants, on farms where later plantings are quite healthy. This favors the theory that blight results from weakness caused by early planting in cold ground. It is a fungous disease, and may sometimes be prevented by the use of Bordeaux mixture. The same remedy is the best known preventive of black rot.

Potato bugs may be either hand-picked or poisoned with Paris green. The tobacco worm sometimes causes much damage to the tomato. All diseased or blighted tomato vines should be promptly burned, and the crop carried to new soil the following year.

Marketing, Profits, etc.—As already stated, choice tomatoes in clean baskets are always in demand, and a new basket will usually pay for itself on a single sale. The sum of $150 per acre may be quoted as the average gross receipts from tomatoes at present prices. This estimate is based on the low yield of a half-peck of fruit to each vine at 25 cents per basket. If sold retail, the tomatoes would command more money, while if sold in bulk to a canning factory the gross receipts might be larger or smaller, depending on the size of the general crop and other circumstances.

TURNIPS AND RUTA BAGAS.

Turnips and ruta bagas are closely related. The latter are turnips in fact, and are frequently called Swedes. The common method on many farms is to sow turnips broadcast, but it is a far better practice to sow both these and the ruta bagas in drills, so that they can be kept clear of weeds and worked by horse-power. Not only are these advantages secured, but the row system makes it possible to take out the superfluous plants, and secure roots of uniform size and shape. Turnips and ruta bagas have high economic value as foods, both for humanity and for live stock.

Turnips.—Turnips are grown for market purposes both in spring and in fall. In the spring the seed should be sown early, in mellow soil. For the fall crop the seed may be sown either in July or August. The rows in garden or field may be as close as can be conveniently worked.

TURNIP.—For earliest, we recommend Purple Top and White Milan. For fall crop, Mammoth Purple Top Globe and Golden Ball. For descriptions, see "Johnson & Stokes' Garden and Farm Manual."

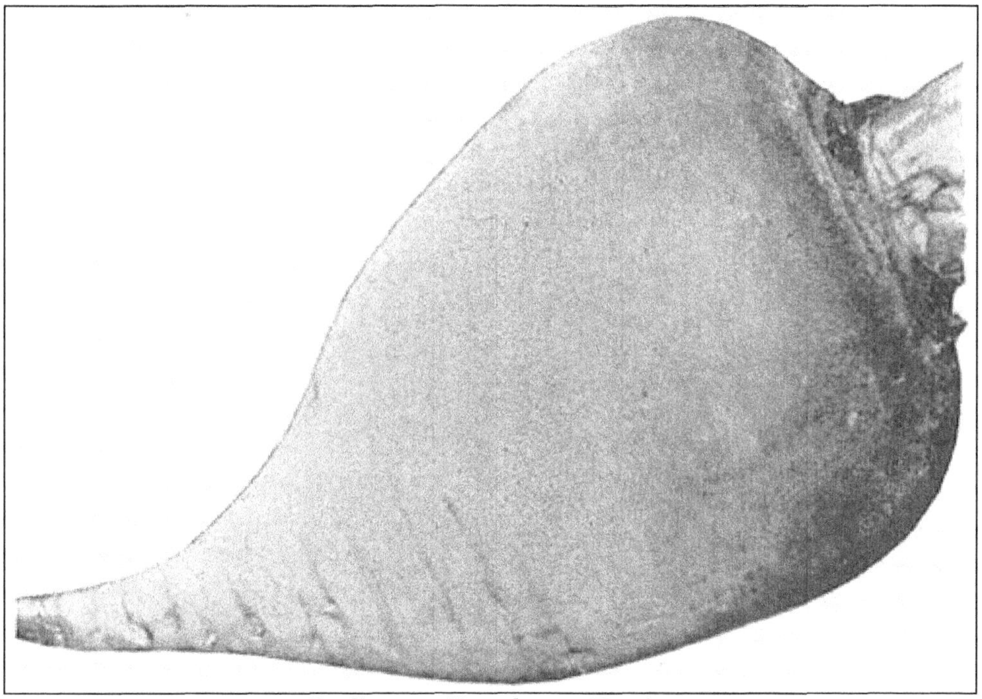

Budlong or Breadstone Turnip.

Ruta Bagas or Swedes.—The seed of ruta baga or Swedish turnip should be sown (in the latitude of Philadelphia) in July, a little earlier than the seed of the common turnip. The ground should be well enriched with rotted manure, the rows 2½ to 3 feet apart, the seed covered to the depth of half an inch, and the plants afterward thinned out so as to stand 6 or 8 inches apart in the row. The crop is almost always large and satisfactory.

Ruta Baga.—We recommend Myer's Purple Top Beauty and Budlong. For descriptions, see our "Garden and Farm Manual."

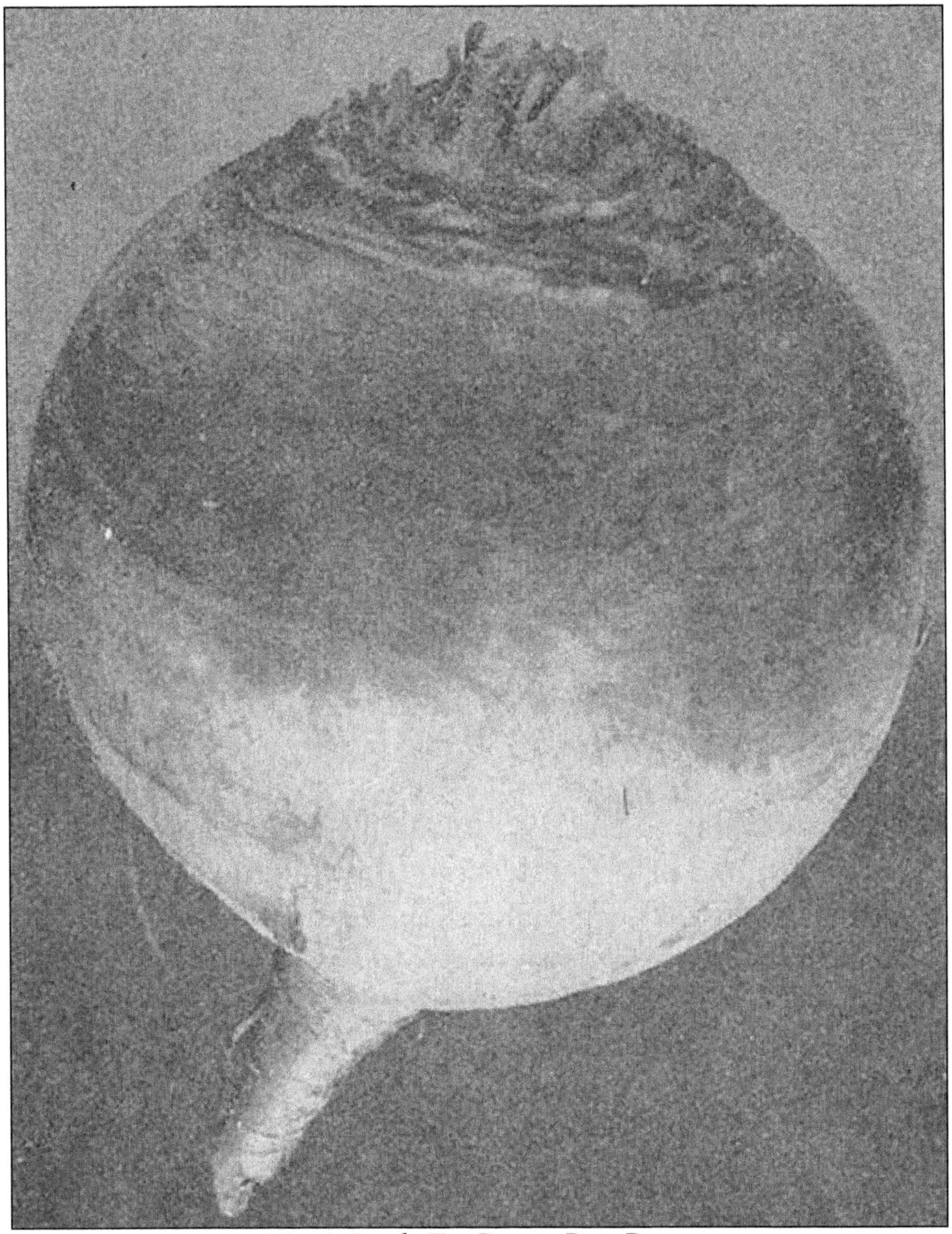

Myer's Purple Top Beauty Ruta Baga.

Storage.—Turnips of all kinds sell well in the winter markets, to say nothing of their high value as stock foods. They are easily preserved in root cellars, covered with sand, or in pits in dry soil, covered with straw and earth to prevent freezing.

CHAPTER IV.

VEGETABLES SUITED TO FARM CULTURE IN SOME LOCATIONS.

In this portion of the book are grouped a number of vegetables not adapted to every farm or location. The list includes celery, water cress, cucumbers, egg plants, kale, lettuce, melons, mushrooms, onions, peas, radishes, rhubarb, spinach, sweet potato, etc. Where favored locations for their production exist on farms they may be grown with profit, if markets are accessible.

CELERY.

On very many farms there are meadows with deep, rich soils that are now lying under grass; or, worse, under tussocks and swamp weeds. Some locations are subject to disastrous overflow during freshets, but innumerable spots exist where such meadows could with safety be converted into celery gardens, capable of easy irrigation, either situated above the level of floods or susceptible of artificial protection by means of cheap embankments. Such situations are entirely too valuable to use for pasturage. They are the truck gardens of the future.

Perfect Celery.—The object in celery-growing is to produce thick, robust, tender, solid, crisp, sweet leaf stalks, free from rust or insect attacks. The essentials are rich land and plenty of water, and skill is required in the two points of bleaching and storing. But there are no mysterious processes to be learned. The Kalamazoo growers have, it is true, a rare advantage in their deep muck soil, with a permanent water level only a few inches or feet below the surface, but their success depends on accuracy of working detail almost as much as on perfection of soil. It is not necessary to go to Michigan for good celery ground.

Fertilizers.—The best known fertilizer for celery is thoroughly rotted barnyard manure. Fresh manure is to be avoided for several reasons. It is less available for plant food, more likely to produce rust, and more liable to open the soil and render it too dry. Commercial fertilizers are not infrequently used, but there is a decided preference among many celery growers for the rotted stable product. Shallow plowing (5 inches) is practiced, as celery roots do not go deep.

Planting.—It requires from 20,000 to 35,000 celery plants to the acre, according to their distances apart. In the intense culture at the great celery centres two crops (and even three crops) of celery are grown upon the land per year, by a system of planting between rows, but in the operations of farm gardeners not more than one crop per season is grown. This may follow an earlier market crop, such as peas, beans, onions or sweet corn, though where the farmer is hard pushed with other work, the celery may be grown without any other crop preceding it, but not upon newly-turned sod land, as the earth should be loose and mellow.

Seed for early celery must be started under glass, but the farmer will find his best celery market in the autumn. April will, therefore, be ample time for sowing the seed, which should be scattered thinly in rows in finely-raked mellow soil in the open ground, and covered lightly. The seed is very slow to germinate, and the bed should be copiously [Pg 77] [Pg 78] watered until the plants are well started. In small operations, it is well to transplant at least once. In large operations, the plants are thinned out in the original rows, and carried from thence direct to the field. The upper leaves and the tips of the roots are cut off, and the plants are set firmly in the soil by means of a dibber.

J. & S. Golder Self-Blanching Celery Prepared for Market.

Dates and Distances.—July is a proper time for setting out celery; preferably after a rain or during dull weather. The rows may be from 3 to 5 feet apart, depending on the purpose of the planter, and the plants 5 or 6 inches apart in the rows. If the celery is to be stored for blanching, 3-feet rows may be used. If it is to be blanched in the field, the distance between the rows should be greater, so that more loose soil will be available for hilling.

One ounce of celery seed will furnish 2,500 to 3,000 plants. A half pound is sufficient to furnish plants for an acre.

Even on good ground celery should not be set out later than August 15th (in the latitude of Philadelphia), and preferably earlier.

The system of level planting is practiced by large growers everywhere. Trenching is still followed in some private gardens, but is too expensive for commercial operations.

Varieties.—The so-called dwarf and half-dwarf varieties have pushed the larger kinds out of the market almost entirely, though seed of the giant sorts can still be obtained. The dwarf kinds are large enough for all purposes, however, and are in best favor everywhere. They are about 18 inches high, as compared to twice that height in the old-fashioned giant types.

The favorites of late years for early celery are the self-blanching sorts, such as White Plume and Golden Self-Blanching, which are the result of the continued selection of individual plants or sports showing a tendency to blanch easily. For winter keeping, the Perle Le Grand, Winter Queen and Perfection Heartwell are the best. These varieties are beautiful as well as highly palatable. There are also red or pink sorts, of high table merit and good keeping qualities.

CELERY.—We recommend Golden Self-Blanching and White Plume for early, Perle Le Grand for both early and late and Winter Queen for late. The latter is the very best keeper. See "Johnson & Stokes' Garden and Farm Manual."

Cultivation.—The proper culture of the celery has already been suggested in the allusion to its need for water and its shallow feeding habits. The surface soil should be highly enriched, the stirring of the soil very shallow, and the water supply copious, either by capillary attraction from below (as at Kalamazoo) or by rainfall or artificial irrigation.

Blanching.—The first step in the process of blanching or bleaching is what is known as handling. This operation consists in grasping all the leaves of a celery plant in one hand, while with the other the soil is drawn together and packed so as to hold the stalks in an upright, compact position. This single operation will fit some of the early-planted sorts for market in the course of two weeks; though a second operation, called hilling, is usually considered desirable, even with the self-blanching sorts. See photograph on first page.

The Kalamazoo growers depend on muck for field blanching, though they also use boards. Muck is merely a dark soil, containing or consisting mainly of vegetable matter. They first "handle," as just described, and about five days later draw 6 inches more of the muck about the celery stalks. Again, three days later, they draw an additional 2 inches about the stalks, and in two weeks from the start the celery is ready for market.

These operations are frequently done by two men working together, one holding the stalks and the other drawing the soil to them. The first operation puts the stalks in an upright, compact position, so that little or no soil can get into the heart of the plant. The second draws about the plant all the soil that will conveniently remain there. The third merely supplements the second, as the hill has had time to become somewhat firm and has settled away a little from the upper leaves.

Boards are used for summer blanching, as they are less heating than soil. Ordinary lumber, free from knot holes, is employed. The boards rest on their edges, one board on each side of the row, the tops being drawn together until within 2½ inches of each other, and the lower edge of the board held in place either by stakes or by soil.

The work of handling or hilling must be done only when the celery is dry and unfrozen. In fact, celery must never be handled when wet (except when preparing it for market), or it will surely be rusted and spoiled.

The same practices of blanching celery as here mentioned in connection with the Kalamazoo operations are in vogue near Philadelphia and other Eastern cities, and are not new. The real reason that Kalamazoo is so celebrated is her possession of that wonderful black muck soil, underlaid with standing water. This has attracted the best celery growers of the country; men who have small places of from one to three acres, and who work out every detail to perfection, employing little labor outside of their own families and concentrating their efforts on the production of perfect celery crops. There are extensive celery growers at Kalamazoo, with tracts of thirty or more acres devoted exclusively to this vegetable, but the majority of the gardens there are small, and much hand-work is done.

Winter Storage.—The art of the winter storage of celery, as practiced by large growers, is not hard to learn. Both at Kalamazoo and here in the Eastern States there are two methods in vogue. One is the use of especially-built houses, and the other is the open-field plan.

Blanching Celery with Boards.

The celery house or "coop" is a low frame structure, half under ground, generally 14 or 16 feet wide, and as long as may be desired. There is a door in one end and a window in the other. The sides, ends and roof are double and filled with sawdust. There are wooden chimneys or ventilators at intervals of 12 feet along the peak of the roof, and sometimes there are glass windows in the roof, provided with wooden shutters. The celery stands upon the floor, which is of loose soil. There is a narrow walk lengthwise in the middle of the building, and boards extending from the central walk to the side walls separate the packed celery into narrow sections. No earth is placed between the celery stalks as they stand. They are, in fact, rooted in the soil of the floor, and are thus able to make the slight growth demanded for complete blanching. The various doors, windows and ventilators make it possible to keep the air fresh and wholesome, and during cold weather a stove may afford heat to the storage room. Artificial heat is not commonly required.

Another method, cheaper and quite as satisfactory, especially on farms or in market gardens, is to trench the celery in the open field. The situation of the trench must be a dry one, where there will be no standing water. The trench must be nearly or quite as deep as the height of the celery, with perpendicular sides, and a foot or less in width. The stalks are set upright in the trench, with all decayed or worthless leaves removed, as closely as they will stand, without soil between them. To keep them in that condition is purely a matter of care. If they are buried deeply and the weather proves warm they will rot. But if the covering be decreased in warm weather and increased in cold weather, the celery can be kept in perfect condition. In private gardens celery is often planted in double rows, a foot apart, and wintered where it grows by covering deeply with soil.

An excellent plan is to make an A-shaped trough of two boards to turn the rain, on top of which a greater or less amount of straw, leaves or litter may be piled, if needed.

Mice sometimes do considerable damage to stored celery, but are more easily controlled in short trenches than in long ones.

Small amounts of celery may be stored in cellars, in boxes a foot wide and a foot deep, with damp sand in the bottom. No soil is needed between the plants. The coolest and darkest part of the cellar is best for storage.

Diseases.—Celery diseases are preventable and insect attacks are few. For blight, kainit is recommended, both in the seed-bed and open field. For rust, the Bordeaux mixture is advised. Hollow-stemmed or pithy celery is the result of poor stock or improper soil, and can be avoided by the use of more manure and more water.

New Process.—The method of growing celery in highly enriched soil, with plants set 6 or 8 inches apart both ways, is quite feasible. The plants stand so close as to blanch each other to some extent, but the system has never attracted general favor. A great deal of water is required. Cultivation is possible only when the plants are small.

Profits.—The use of celery is obviously on the increase, but the demand is for a first-class article. The cash results may be set at anywhere from $200 to $500 per acre. The actual net profits of well-conducted operations are considerable.

WATER CRESS.

Water cress, a vegetable closely allied to several other edible cresses, is used in very large quantities in all city restaurants. It is a much-esteemed winter relish, and is mostly served with every one of the thousands of beefsteak orders daily filled in the great eating houses and lunch rooms. The demand for it seems to be on the increase.

Water Cress.

Water cress is of the easiest culture. It can be grown in the soil of a forcing house under glass, and is extensively produced in this way by market gardeners.

The cheapest method is to grow it in running water, preferably near a spring head; and many such situations are available to farmers. Flat beds, made of loam, gravel, or sand, covered with 3 or 4 inches of warm, spring water, will yield great quantities of water cress in early spring; and the use of a few sash will keep the cress in growth during the winter. The cress should be cut frequently, as the young shoots are most succulent and tender.

For market purposes the water cress is tied in bunches, and retailed at from 3 to 10 cents per bunch, or packed in pint boxes, leaves uppermost, and retailed for about 10 cents per box. These are winter and early spring prices. Water cress culture is profitable in favored locations.

CUCUMBER.

The cucumber market is not easily over-supplied, but the pickling tub should stand ready to receive all cucumbers not sold in a fresh condition.

For field culture, good ground must be selected, and marked out with a plow, 4 × 4 feet; or, a little wider, if the soil is strong. At least one shovelful of well-rotted manure is dropped in every hill, and mixed with the soil, and a dozen seeds planted, to be thinned out finally to three or four plants. It is better to have extra plants, on account of the attacks of the striped beetle.

The cucumber belongs to a botanic family which is naturally tender, and the seeds should not be sown until the soil is quite warm. For farm work, the planting season is the latter part of May and the whole of June; and even July is a suitable month, if the soil can be irrigated. It will require two pounds of seed for an acre.

The variety sown should depend on the purpose in view; but in all commercial operations, well-known and thoroughly tested sorts should be chosen. Shallow cultivation is recommended.

If an early market is to be supplied with cucumbers, the seeds may be started under glass, on bits of inverted sod or in small boxes, and set in the open ground on the arrival of settled warm weather; but the farmer will usually find it most profitable to sow the seeds where the plants are to remain.

The most serious enemy of the cucumber vine is the striped beetle, which attacks the young plant and frequently ruins it. The remedy is air-slaked lime, or soot, or sifted coal ashes, or wood ashes diluted with dry road dust. The best preventive is salt or kainit, used in the hills. The true plan is to have strong, vigorous plants, which, as a rule, will resist and outgrow the striped beetle, and be not greatly injured by its attacks. There is a blight which sometimes destroys the cucumber vine, apparently the result of weakness following a prolonged drouth.

The vine of the cucumber must be kept in vigorous growth, not only by cultivation and a sufficient water-supply, but by care in removing all the fruit as soon as formed, for, [Pg 87]
[Pg 88] if the seeds be permitted to mature, the vine will quickly perish. It is the purpose of the vine's existence to produce ripe seeds, and it will make repeated and long-continued efforts to accomplish this end. In gathering the cucumbers, it is important to avoid injuring the vine. Some growers use a knife; others break the stem by a dexterous twist, without injuring the vine in the least.

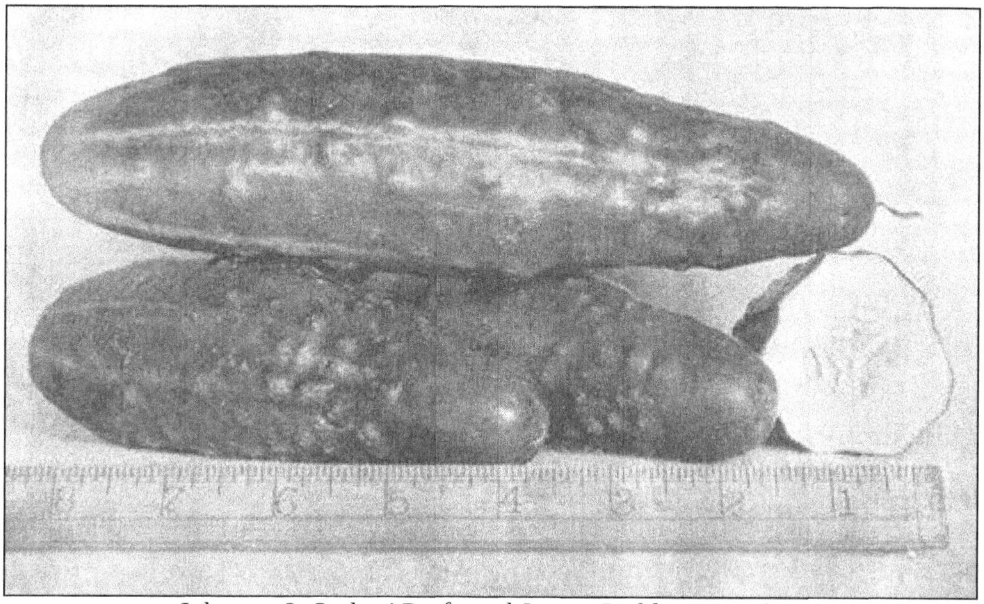

Johnson & Stokes' Perfected Jersey Pickle Cucumber.

It requires 300 cucumbers (more or less) of fair pickling size to make a bushel, and it is estimated that an acre will produce from 100 to 200 bushels, or even more. When the pickles are pulled while quite small, the number runs up to 125,000 per acre; and the pickle factories in some cases make their estimates on a yield of 75,000 per acre. The price is variable, but often quite profitable.

CUCUMBER.—For planting in the South to ship to Northern markets use Improved Arlington White Spine. Giant of Pera is a fine table sort. For pickling, plant Johnson & Stokes' Perfected Jersey Pickle. For description see our "Garden and Farm Manual."

Downy Mildew.—A disease which lately threatened to destroy the business of growing pickles in New Jersey and elsewhere, the downy mildew of the cucumber, can be fully overcome by spraying the vines with Bordeaux mixture. It requires six or seven applications, at intervals of a week or ten days, to conquer this comparatively new disease. Downy mildew is a fungous trouble affecting the leaves and destroying the further usefulness of the vine. A recent New York experiment showed a yield of $173 worth of pickles per acre under spraying as against complete failure where the Bordeaux mixture was not used. The cost of spraying was $9.50 per acre, leaving $163.50 per acre as the value of the crop saved by the operation.

EGG PLANT.

The advisability of growing egg plants in farm gardening operations is a question of location. On a suitable soil, near a good market, the operation will be a profitable one, if rightly managed. The egg plant is a tender vegetable, botanically allied to both the tomato and the potato, but less hardy than either, especially when young. For this reason it is best to delay sowing the seed, even in hot-beds, until cold weather is past, for the tender seedlings never fully recover from a chill or set-back. Indeed, for the farm gardener the month of May is early enough to sow the seed under glass, for this plant grows with great rapidity in a warm soil, and May-sown seed not infrequently yields plants that outstrip those sown a full month earlier.

New Jersey Improved Large Purple Smooth Stem Egg Plant.

The egg plant demands a richer soil than either the potato or tomato. It also asks for more water. It is a rank feeder. A good stimulant, if rotted manure cannot be had, is nitrate of soda at the rate of 400 pounds to the acre.

The farm gardener will do well to consider his market before engaging in the production of the egg plant on an extensive scale, for it is a perishable product. It bears shipment well, but its use is mainly limited to consumption while fresh. It may command a very high price at some seasons of the year and at other times be practically unsalable at any price, owing to an over-supply.

If egg-plant seed be sown under glass in early May, and carefully protected against cool weather (especially at night), the young plants will be ready to transplant before the end of the month and large enough for the open field in June. They should be set in rows 4 feet apart, and about 3 feet apart in the row. Set at these distances, an acre of ground would accommodate about 3,500 plants.

The enemy of the egg plant, in growth, is the potato bug, which must be hand-picked or poisoned. There is a rot which causes the fruit to drop from the stem before reaching maturity. This rot is a fungus, and the Bordeaux mixture is recommended for it. The blight which sometimes affects the foliage is in part at least caused by cold weather, and for this there is no remedy, except late planting.

Every healthy plant should produce from two to six or more full-sized fruits, and it is therefore easy to calculate that an acre's product under favorable circumstances may be very large.

Egg Plant.—There is nothing equal to the New Jersey Improved Large Purple Smooth Stem for the use of farm gardeners. For description, see "Johnson & Stokes' Garden and Farm Manual."

KALE OR BORECOLE.

Kale, of which there are many varieties, is a headless cabbage, closely allied to such vegetables as Brussels sprouts, collards, etc. It is one of the most hardy of vegetables, and in this latitude it will live over winter in the open ground, with only straw or litter as a protection. If cut for use when frozen it should be thawed out in cold water. The kales are among the most delicately flavored cabbages. Some of them are of such ornamental shape as to be full worthy of cultivation for decorative purposes. The height varies from 1 to 2 feet, and the colors include both greens, dark purples and intermediate shades.

Kale demands a rich, deep soil. The seed should be sown in a border or seed-bed, and transplanted to the open field and set in rows, after the manner of cabbage. It is largely and profitably grown in the South for shipment to the great Northern markets. Where farmers are situated near centres of population where kale is in demand, its culture will be found profitable, as it requires even less labor than cabbage. It is planted both in spring and autumn. The former crop is for autumn consumption and the latter crop is carried over winter after the manner of spinach, protected by a light covering of some sort of litter.

KALE.—For the South, we recommend Extra Dwarf Green Curled Scotch; for the North, Johnson & Stokes' New Imperial. See our "Garden and Farm Manual" for descriptions.

LETTUCE.

In some sections, especially in the South, lettuce can be grown with profit by farm gardeners. Depending on the latitude, the seed may be planted from autumn until spring. The plants are usually sheltered and headed under glass, or under muslin-covered sash, and are sent North in ventilated barrels.

The lettuce is naturally a cool-weather plant, and its culture [Pg 93] [Pg 94] is easy. The seed is cheap and it germinates quickly. Well-grown lettuce always commands good prices. It is usual to start the seeds in a border or under a frame, and to prick out the plants into more roomy quarters as soon as they are large enough to handle. In a few weeks after transplanting, in good growing weather, they are headed ready for market. Good soil, abundance of moisture and free ventilation are essentials in lettuce production.

In some parts of the North lettuce culture would be found profitable by farmers in the summer season, for there are varieties well adapted to high temperature, provided good soil and sufficient water be furnished. There is not a month in the year when lettuce is not demanded for use in salads, and this demand is likely to increase.

LETTUCE.—For the South, we especially recommend Reichner's Early White Butter, Big Boston and New Treasure; for the North, New Sensation, Mammoth Salamander and Hornberger's Dutch Butter. Please see "Johnson & Stokes' Garden and Farm Manual."

MELONS.

Melon culture belongs on the farm rather than in the small market garden, on account of the large space occupied by the growing vines. An acre of ground will accommodate only about 450 watermelon hills (at 10 feet each way) or about 1,200 muskmelon or cantaloupe hills (6 feet each way), and hence the necessity for large areas of ground for the cultivation of these crops.

The requirements of the various melons are quite similar. Broken sod ground or any green crop turned down favors their growth, and well-rotted stable manure in the hill is the best known stimulant. All the melons are tender, and are suited only to warm-weather growth, and this fact must be remembered in sowing the seed. Light alluvial soil near rivers or streams is adapted to melon growth, and many an old meadow now weedy and unprofitable might be used to advantage for one of these crops.

New Black-eyed Susan Watermelon.

The Watermelon.—For cash-producing purposes the best watermelon is a large one, with a hard rind. It must have a dark pink or red centre and must be a good shipper. It should weigh thirty to forty pounds, and there should be 900 to 1,000 first-class melons to the acre.

The best melon for family use or for a strictly retail trade is a medium-sized variety, which has a thin rind, pink or red flesh and extra sweetness, weighing from twenty to thirty pounds.

The preparation of the ground has already been suggested. Two shovels of manure should go into each hill. The planting date is May in this latitude; or as soon as the ground is thoroughly warm. Four pounds of seed per acre [Pg 96] [Pg 97] will be required. But one plant per hill is allowed to grow. The end of the main shoots should be pinched off, to encourage branching and flowering.

Cultivation should be thorough. Fungous diseases can be controlled by means of the Bordeaux mixture, except that it is difficult to reach the under side of the leaves. To prevent sunburn on melons, some growers sow buckwheat when the vines are in blossom, and thus secure a partial shade by the time the fruit is large enough to be injured by the sun. Generally, no protection is necessary.

At $10 or $15 per hundred, the average wholesale price at Philadelphia, watermelon culture is profitable. Early prices are higher.

WATER MELONS.—For shipping—Johnson's Dixie, Blue Gem, Duke Jones, Sweet Heart. For home market—Black-Eyed Susan, Florida Favorite, Kentucky Wonder, McIver's Wonderful Sugar. For descriptions, see "Johnson & Stokes' Garden and Farm Manual."

Citron.—This small round melon is cultivated in all respects as the watermelon, but being smaller the hills may be closer. It is used in making preserves. The name citron is frequently applied to certain of the cantaloupes.

Cantaloupes or Muskmelons.—It is a matter of choice whether the green-fleshed or red-fleshed sorts are grown; or whether the variety be large or small. The sorts covered with strongly webbed or netted markings are in high favor for shipping to distant points, as they carry well. Flavor is in part at least a matter of temperature and sunshine. Cantaloupes may be nicely ripened by removing them from the vines and storing in dry, warm rooms.

The usual planting distance is from 4½ to 6 feet, in hills containing rotted manure. Compost, made of hen manure, is sometimes used in the hill, well mixed with the soil. Good cantaloupes are always in active demand.

Musk Melons.—Early sorts for shipping—McCleary's Improved Jenny Lind, Netted Beauty, The Captain, Champion Market, Improved Netted Gem, Anne Arundel. Late sorts—The Princess, Johnson & Stokes' Superb, etc. See "Garden and Farm Manual" for descriptions.

McCleary's Improved Early Jenny Lind Muskmelon.

Enemies.—In addition to the fungous diseases of the watermelon and cantaloupe, which are best treated with Bordeaux mixture, all melons are sometimes badly troubled with an aphis called the melon louse. The remedy is whale-oil soap—a pound in six gallons of water; or kerosene emulsion. The latter is made by dissolving half a pound of soft soap in one gallon of water; then adding two gallons of kerosene, churning violently; then diluting with ten or twelve gallons of water. This emulsion is put upon the melon vines in the form of a spray, and is one of the best insecticides known. It is to be used on all sucking insects, like lice and squash bugs. Biting insects are easily killed with Paris green—one pound in 100 pounds of flour or plaster, or in 150 gallons of water.

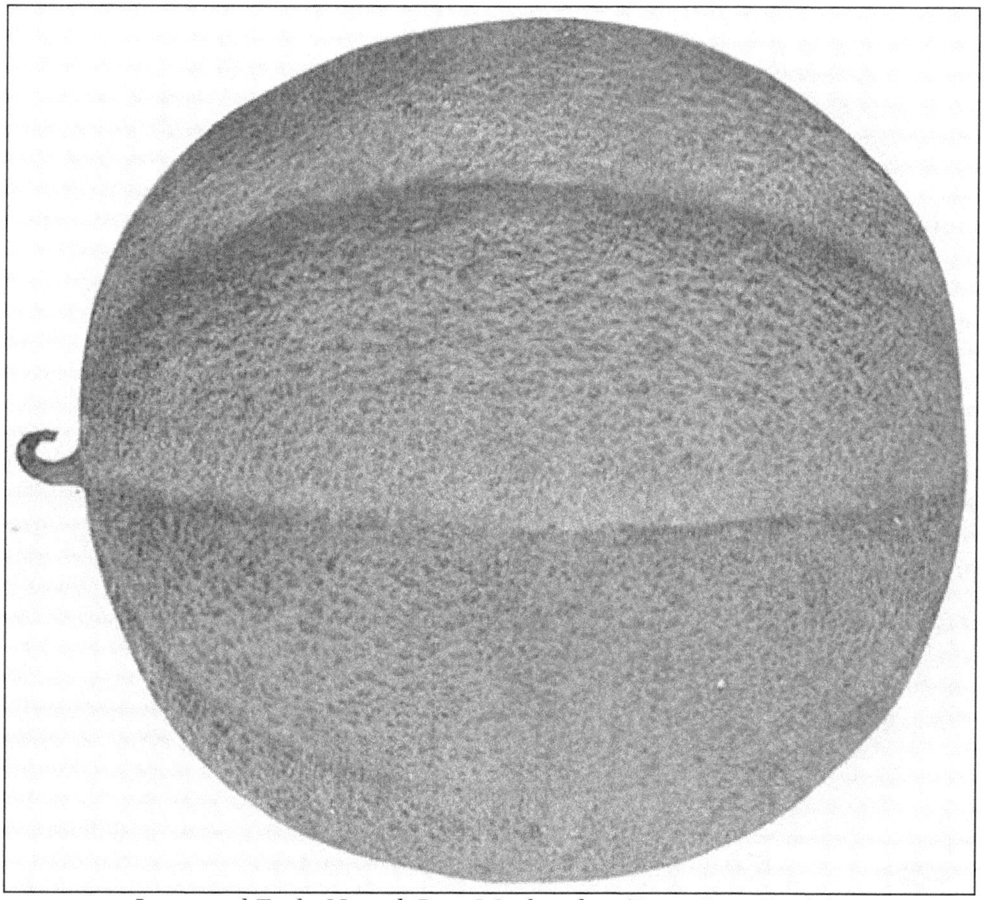

Improved Early Netted Gem Muskmelon (Rose Gem Strain).

Where the land is suited to melon culture, in any part of the country, the farm gardener will find no more satisfactory or remunerative crop.

MUSHROOMS.

Under certain favored circumstances the mushroom may be grown as a farm gardener's crop. The requisites are horse manure and a dark cellar, cave or vault. If the manure be available and a suitable apartment at hand, the growing of mushrooms may be taken up for winter work.

A Bed of Mushrooms from English Milltrack Spawn.

There are many ways of growing mushrooms, and they can be produced in any situation where a steady temperature of 60° can be maintained. A simple method is to prepare a bed consisting of horse manure and loam, three parts by measure of the former and one of the latter, the manure having been somewhat fermented and sweetened by allowing it to heat and turning it several times. A compact bed a foot deep is made. This bed will first heat and then cool. As it cools, when at 80° or 85° an inch below the surface, bits of brick spawn the size of a hen's egg are inserted about 9 inches apart.

The bed must not be immediately covered, or the temperature will rise sufficiently to kill the spawn. In ten days, more or less, as shown by a thermometer, this danger will be past, and the bed should receive a coating of good loam an inch deep. No water is to be applied until after the bed is in full bearing.

It is assumed that the temperature of the room or cellar has been uniformly 60°, day and night; that the bed has not been made where it could become water-soaked; that it is sufficiently moist, yet not wet; and that no draft of air has passed over the surface in a way either to reduce the temperature of the bed itself or to dry the soil upon the surface. If these conditions cannot be maintained, either by a specially favorable place or by means of covering the bed with litter, it is better to let mushrooms alone.

The crop should appear in six or eight weeks, and should last two months, the total product being from one-half to one pound per square foot. The cash price is from 50 to 75 cents per pound in the large cities; and the crop is sufficiently profitable to warrant the losses which beginners so commonly experience. These losses are the result of carelessness or ignorance in the matter of details.

The usual sources of failure are poorly prepared beds, the medium being either too wet or too dry; frequent changes of temperature; improper use of water; and, lastly, poor or stale spawn.

Mushrooms are packed in small baskets lined with paper, and carefully covered to prevent evaporation. A five-pound package is a favorite shipping size.

ONIONS.

The onion is a national crop; as widely though not quite as extensively grown as the potato. It is available as a money crop for the farm gardener.

Choice of Soil.—Heavy, stiff clay land is to be avoided. Sand and gravel dry out too quickly. Stony land renders good culture difficult. The best soil for onions is a deep, rich, mellow loam. Soils which afford natural advantages for irrigation should not be overlooked, as the rainfall is often lacking when greatly needed.

Fertilizers.—Onion culture demands high manuring. No amount of rotted stable manure is likely to be excessive. A ton per acre of high-grade, complete fertilizer is not too much, if moisture can be supplied. Hen manure is a good top dressing for onion-beds, furnishing the needed nitrogen. Nitrate of soda is a good source of nitrogen, if nitrogen must be purchased. The clovers and other leguminous crops yield the cheapest nitrogen. Wood ashes, kainit, etc., furnish potash. Either ground bone or acid phosphate will give the needed phosphoric acid. An analysis of the onion shows that it carries away fertility in just about the proportions furnished by stable manure.

It is a singular fact that onions can be grown year after year on the same ground, if well manured. Rotation is necessary only in case of the occurrence of disease or insect attack. The onion loves cool weather.

Planting.—To grow onion sets, the seed is sown in close rows, at the rate of from fifty to sixty pounds per acre. To grow large onions direct from seed, five pounds of seed per acre will be required. To plant a field with onion sets will require twelve to fifteen bushels per acre, according to size of the set.

A List of the most Popular American Onions.

An onion set is merely an immature bulb. Sets vary from the size of a large pea up to that of a walnut. When the seed is sown thickly the bulbs have no chance to grow, and the summer weather quickly ripens the tops, completely suspending the growth of the bulb. In some parts of the country onion sets cannot be grown with profit, as the tops refuse to die and the bulbs or sets do not ripen properly.

In nearly all parts of the United States onions can be grown direct from the seed the first year; especially from seed grown around Philadelphia, which is earlier than Western-grown. It is quite customary in the South to sow onion seed in late summer or autumn; in August or September. This will give early spring onions of marketable size. In the North, within quite recent years, it has become the practice to sow onion seed in frames, in fall or early spring, and transplant the young onions to the open ground. This is sometimes spoken of as the new onion culture.

Onion sets or young plants should be placed 3 or 4 inches apart, in rows a foot apart, if to be cultivated by hand; the rows farther apart if for horse work.

The onion is hardy. Many varieties will live in the open ground over winter, if covered (at the North) with light litter. It is in this way that shoots for bunching are obtained early in the spring.

The seed should be sown for sets when the apple is in bloom. Sets may be put into the ground earlier; in fact, as soon as the ground can be worked. The set should not produce seed the first year, though it often does so. It should, on the contrary, grow to the size of say 3 inches, and then ripen for winter storage. Excessively large onions are not desirable. To hasten maturity, the tops may be broken down or the roots may be cut by running a knife or sharp plow or cultivator along one side of the row.

The onion, under favorable circumstances, will produce [Pg 105] [Pg 106] a crop of 800 bushels (fifty-six pounds to the bushel) per acre; though 500 bushels is nearer the average product.

Weeding a Field of Onion Sets on our Bucks County Seed Farm near Philadelphia.

Storage.—The storage of onions and onion sets is simple. The bulbs should first be ripened on the ground, by a brief exposure to wind and sun. This completes the wilting of the tops. They should then be spread out on ventilated trays or racks, or a few inches in depth on a floor, in a dry, shady place, where the air is good, preferably a loft; not a damp cellar. Freezing will not injure them, but they must not be handled when they begin to thaw, or they will rot. They must not be bruised during the operation of gathering or during the process of storage.

A popular and excellent method of wintering onions in cold climates is to spread straw to the depth of 18 inches on a dry floor or scaffold, and put on a layer of onions from 6 inches to a foot deep, and cover with 2 feet of straw. This will not always prevent freezing, but it checks all sudden changes.

Onions not fully cured should never be kept in barrels, but spread out so as to be perfectly ventilated. Onion sets shrink greatly in storage; sometimes as much as one-half between fall and spring.

Varieties.—There are many varieties of onions, some of American and some of foreign origin. The former are better keepers, but the latter are of milder flavor. The American sorts (Danvers, Southport Globe varieties, Wethersfield, Extra Early Red, Silver Skin, Strasburg, etc.) are usually considered to be the most profitable; but the foreign kinds (Prize Taker, Prize Winner, Pearl, Bermuda, Giant Rocca, Victoria, etc.) are profitable in those parts of the country where soil and climate warrant their growth from seed in a single season.

The so-called tree onion is a perennial, of American origin, living out over winter. It is sometimes called Egyptian or top onion. It produces bulbs or sets at the top of the seed-stalk.

The potato or multiplier onion divides its large bulb into numerous small ones, which in turn produce large onions the next year.

ONIONS.—For farm gardeners' purposes, we especially recommend Philadelphia Yellow Globe Danvers, Mammoth Yellow Prizetaker, White Prize Winner. Earliest Onions are—Extra Early Red Globe Danvers, American Extra Early White Pearl, Rhode Island Yellow Cracker. The best for sets—Extra Early Red, Philadelphia Yellow Dutch and White Silver Skin. For descriptions of varieties, see "Johnson & Stokes' Garden and Farm Manual."

Diseases and Enemies.—To prevent maggot, the use of kainit is recommended; 600 pounds per acre. For onion smut, which may in part be cured by the kainit, the best known remedy is a change of soil. Thrip, which causes the cuticle of the leaves to become covered with whitish or yellowish spots, is best treated by means of kerosene emulsion, used as a spray. The onion fly may, in part, at least, be abated by the use of equal parts of wood ashes and land plaster dusted very thoroughly on the young plants. Stiff-necked onions, often called stags, are the result either of improper growth or poor stock. They are sometimes planted in autumn for use as scallions (scullions) the following spring.

Marketing.—Onions are sometimes sold in the open field; a good plan when a fair price can be secured. After curing, as already described, they are usually sold by the bushel or barrel. They are always in demand, as the onion is a standard article of human food.

In the green state they are sold either by measure, by the bunch, or by the rope. The latter method consists in tying the onions along wisps of straw.

Scallions.—No small amount of money is expended by housekeepers in the early spring markets for scallions (scullions), or bunched onion shoots. These tender shoots are washed, tied and sold for 3 to 5 cents per bunch, retail, or half those figures wholesale. Scallions are produced from either sets or large onions planted the preceding autumn, and sheltered either by frames or litter, so as to encourage early spring growth.

PEAS.

It will require one and one-half to two bushels of peas to seed an acre, and no crop finds a more ready sale than fresh peas in the summer and autumn markets. Farmers who are near centres of population, or who enjoy good shipping facilities, will find peas a quick money crop.

Any good soil will produce a crop of this excellent vegetable, but it must not be assumed because the pea is a legume, with nitrogen-collecting roots, that it will not well repay the application of manure to the soil. Peas and beans need less assistance than some other things, but they give good returns for the application of rotted manure or artificial fertilizer.

The seed should be put into the ground in early spring, as soon as the soil is dry enough to receive it, beginning with the smooth, extra-early sorts, which are more hardy than the wrinkled varieties. A little subsequent frost will do no harm.

The smooth, early sorts should be sown in rows, about 3 feet apart, the intermediate or half-dwarf sorts in rows 4 feet apart, and the tall, late varieties, in rows 5 feet apart.

In field operations no sticks are used, and large pickings are taken even from the tall-growing vines while sprawling upon the ground; and the labor is vastly less where no sticks are employed.

The early peas should stand closer in the rows than the later and larger sorts. The Extra Early kinds mature in fifty to fifty-five days from germination; the intermediate kinds in sixty-five to seventy days, and the tall and late kinds in seventy-five to eighty days. For autumn planting, the extra early varieties are used, and are planted until sixty days before frost.

Mildew is a field enemy of the pea, resulting from unfavorable weather. The weevil often attacks the seed, but does not injure it for market purposes.

The canning of green peas is now an industry of enormous extent in America. The peas are shelled and sorted by machinery, and thousands of bushels are annually disposed of in this manner.

The wholesale market price of peas in the pod varies from 50 cents to $3 per bushel at Philadelphia. The latter price is for the early product. The usual retail price is 15 to 25 cents per half peck. The crop of green pods per acre may be rated at 100 bushels, more or less.

PEAS.—Earliest for the South—Johnson & Stokes' New Record Extra Early, Alaska; second early—Johnson & Stokes' Second Early Market Garden; late—Giant Podded Marrow, Improved Stratagem, Crown Prince, Sugar Marrow. For descriptions, see "Johnson & Stokes' Garden and Farm Manual."

RHUBARB.

In some parts of the United States rhubarb or pie plant is grown in very considerable quantities for market purposes, and with profit. Its culture is extremely simple. It is merely necessary to plant seed or roots, and to have the plants about 4 feet apart each way in a permanent bed. The plant is a perennial, lasting for many years. It is a rank feeder, and the more manure given it, the larger and more succulent will be the young shoots. The roots should be divided every five years, as they finally become too large. The demand for rhubarb continues through the spring and into summer, and large quantities are canned for pie-making. Five leaf stalks make a large bunch. It is worth $2 to $3 per 100 bunches, wholesale.

RADISH.

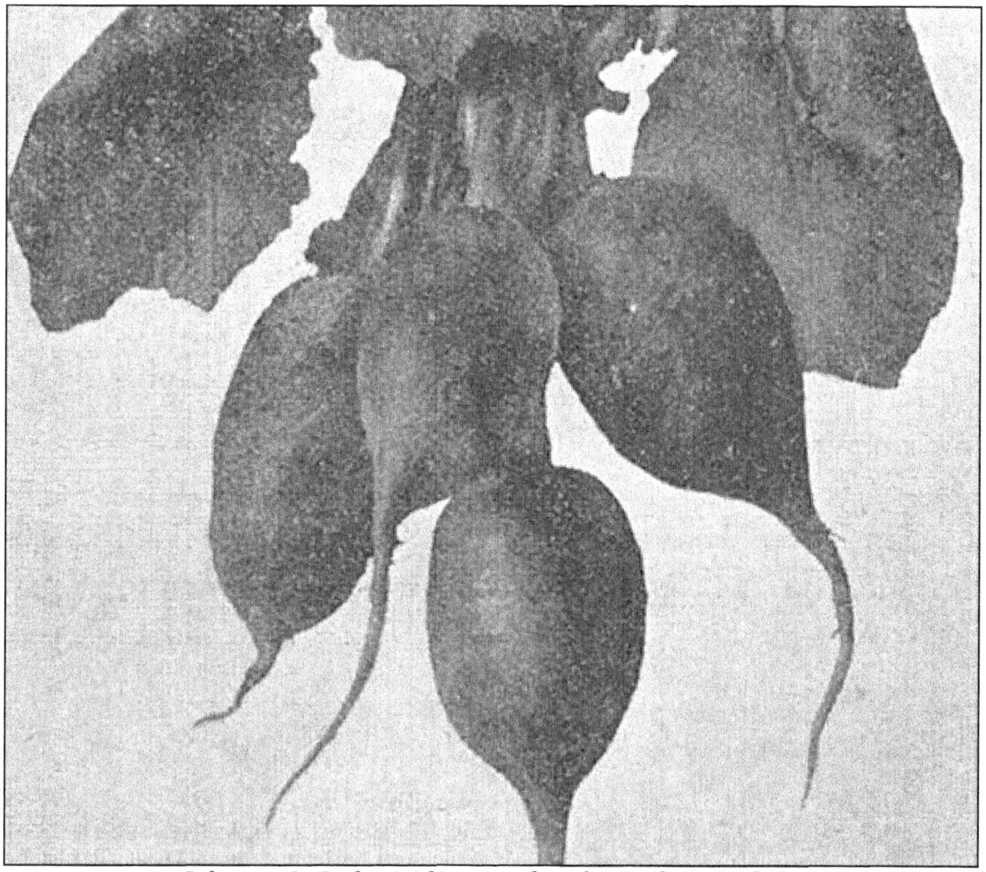

Johnson & Stokes' Olive Scarlet, the Earliest Radish.

Farmers who retail their produce should raise radishes. Rich ground and abundant moisture are the requisites for quick growth, and upon quick growth depends good quality. Slow-growing radishes are hot and pithy. The early sorts are best for spring, but the so-called summer radishes are best for warm weather, as they are not so liable as the early kinds to become pithy. Enormous quantities of winter radishes are grown in autumn, for use and sale during the winter months. They are kept in sand, like other roots.

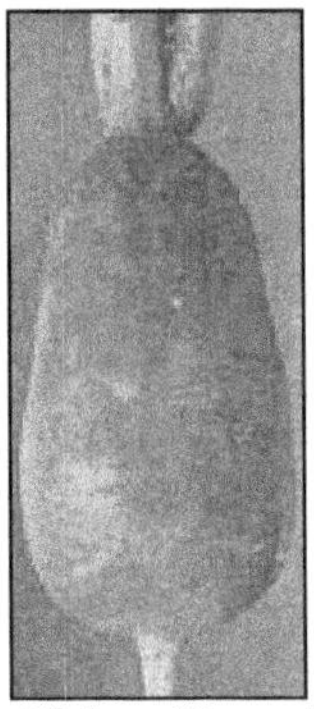

China Rose Winter Radish.

The early kinds mature in twenty to twenty-five days from sowing. Nitrate of soda in small quantities is one of the best known stimulants. Rotted stable manure is good, but hog manure and night soil are not in favor among radish growers, tending to produce insect attacks. The free use of lime, salt or kainit is recommended as a preventive against insects. Sometimes it is necessary to avoid manure of any kind, on account of maggots, depending wholly on artificial fertilizers. As a last resort the radish-bed must be removed to new ground, as the maggot renders radishes wholly unsalable.

The green seed pods of radishes are sometimes used for pickling. The plant is closely related to the mustard.

It is wrong to wait for radishes to grow large (except the winter sorts), as they are sweetest and most succulent when comparatively small. Crisp, sweet radishes always command ready money.

Radish.—Early, for the South—Scarlet Turnip White Tipped, Johnson & Stokes' Olive Scarlet, Philadelphia Gardeners' Long Scarlet. Summer radishes—Red and White Chartier, White Strasburg, Improved Yellow Summer Turnip. All seasons, radishes which are equally good for summer or winter—New Celestial, New Round Scarlet China. For winter use only —China Rose. For descriptions, see "Johnson & Stokes' Garden and Farm Manual."

SPINACH.

Spinach (or spinage) is grown for its leaves, which are cooked in winter and spring for use as "greens." The leaf is sweet and palatable even when raw, but it is always stewed for table purposes. It is a cool weather plant, almost perfectly hardy. It may be sown in spring, for immediate use, or in the autumn for fall cutting, or for carrying over winter.

Plants and Roots of Parisian Long Standing Spinach.

It is of the easiest culture, requiring ten or twelve pounds of seed per acre, either broadcasted or sown in rows. In small gardens it is usually grown in rows, but in open field culture it is more commonly broadcasted. Patches of many acres in extent are seen near the large cities. It is also grown quite extensively in some parts of the South for shipment to Northern markets during January and February.

To prepare it for market the leaves are cut before the seed stalk appears, and after washing are barrelled or crated for shipment. Growers receive from $1.50 to $2.50 per barrel in Philadelphia and New York in the winter and spring. Where accessible to market, spinach is a profitable crop.

Blight is the main enemy. The remedy is removal to another soil.

Of spinach there are many types; some smooth and some with savoy or wrinkled leaves. The property of standing a long time before going to seed is desirable, especially when sown in the spring, as it increases the length of the cutting season.

At the North a slight protection of litter or straw is necessary in winter. South of latitude of Washington no protection is needed. Spinach is cut even when frozen; in fact, at any time when there is no snow on the ground. By throwing it into cold water it quickly thaws, and affords a palatable and healthful food in midwinter. The dead or yellow leaves should be removed before sending it to market, and if carefully prepared it has an attractive green appearance during cold weather when other vegetables are scarce. The winter crop is larger than any other, but much is also grown for spring sales. It is admirably adapted to farm culture.

SPINACH.—For spring planting, we recommend Parisian Long Standing; for autumn, American Savoy or Bloomsdale. See "Johnson & Stokes' Garden and Farm Manual."

THE SWEET POTATO.

The cultivation of the sweet potato affords profitable employment to thousands of American farmers. It is pre-eminently a farmer's crop, on account of the ground space occupied. It demands a light or sandy soil, well drained and well manured. It has wonderful drouth-resisting qualities; though, on the other hand, it is quite unable to withstand continued cold, wet weather. Its territorial range may be said to include nearly the whole of the United States, where the soil is suited to its growth, and it is even cultivated in Canada. It will in all probability increase in favor as it is better known and the manner of preserving or storing it is better understood.

SWEET POTATO.—We recommend and endorse the Hardy Bush or Vineless Sweet Potato. For description, see "Johnson & Stokes' Garden and Farm Manual."

Fertilizers.—There is wide diversity of practice in the matter of enriching the land for sweet potatoes, and most of the standard manures are used, either in one place or another. There seems to be an almost universal endorsement of well-rotted stable manure, and next in favor is wood ashes. High-grade fertilizer of any kind, thoroughly incorporated with the soil, may be used.

Young Plants.—Sweet potatoes are propagated by sprouts obtained by laying tubers on their sides, not touching each other, covered with soil, in specially prepared heated beds. These sprouts produce abundant rootlets while still attached to the parent tuber, and by pulling them with care, great numbers of young plants can be obtained. A second and even a third crop of young plants may be pulled from the same tubers. In the South no artificial heat is needed.

Growing the Slips or Sprouts as Practised in New Jersey.—The fire-bed, so-called, is quite generally used in Southern New Jersey for obtaining slips or sprouts for spring planting. It is necessary to have bottom heat and a uniform temperature of about 70°.

Plant of New Hardy Bush or Vineless Sweet Potato.

The fire-bed consists essentially of a pit about 15 by 50 feet in size. It is floored with boards laid upon cross pieces. Beneath the boards there is an air chamber. On top of the boards the bed is made. At one end is a furnace, with flues running out into the air space beneath the bed, but not reaching the chimney or smoke-pipe at the opposite end of the bed. At the hottest end of the bed the soil is over 6 inches deep. At the cool end a depth of 6 inches is quite sufficient. The whole bed is covered either with canvas muslin or with glass sashes, there being a ridge pole above the bed, running lengthwise with it, thus giving a double pitch to muslin or to glass.

After the soil has been heated somewhat, the tubers are laid on the bed, about an inch apart, and covered with about 3 inches of good soil, and the soil, in turn, covered with leaves or hay, to increase the warmth of the bed. In a week, more or less, the sprouts will show above the surface of the soil, when the leaves or hay must be removed.

The object in not connecting the flues from the furnace with the chimney is to economize heat. The air chamber under the entire bed becomes evenly heated, and the smoke escapes finally by the chimney. This chimney may be made of wood, and a height of 8 or 10 feet will afford ample draft. Either wood or coal may be burned, but preferably wood.

The planting distance in the field is about 3 feet by 2, the young plants being set upon ridges. It requires about 9,000 plants to the acre. The work must not be done until the ground is warm. The crop is ready in from sixty to ninety days.

Cultivation.—Shallow cultivation is all that is required. The vines at the North are not permitted to take root along their length, but in the South they are sometimes allowed to do so, and additional tubers thus secured. At the North the vines are lifted and turned, to clear the way for the cultivator and to prevent rooting.

Enemies.—Black rot is one of the worst of sweet potato diseases. Stem rot is another serious enemy. The best treatment for these and other fungous troubles is prevention, and the best prevention is a healthy soil. It is, therefore, best to go to new land occasionally.

Harvesting.—The common practice is to plow the sweet potatoes out of the ground just after the first frost has touched the vines. The tubers must be exposed to the air for a time, and partially dried. They are prepared for market, if wanted immediately, by rubbing off the soil and sorting into two sizes.

Storage.—At the South one of the several methods of winter storage is to build a light wooden flue of lattice work, and pack about it a conical-shaped heap containing about forty or fifty bushels of sweet potatoes. Straw is used as a covering, with earth upon the straw, the earth to be increased as the weather becomes colder. Over the entire heap a rough shed is erected to turn the rain. The top of the flue or ventilator is closed with straw in really cold weather. The spot must be a dry one.

The New Jersey sweet potato house is a stone building, say 16 × 18 feet on the inside, with walls 10 feet high, and a good roof. The building is half under ground, and the earth is banked up around it. There is a passage way through the centre, and the bins for the sweet potatoes are 6 to 8 feet square and 8 to 10 feet deep. There is a door on the south side, with window above, and a stove is placed inside the building, for use when required. The walls are plastered, and the under side of the roof is also covered with lath and plaster, and the place is thoroughly weather-proof. A house of this kind will afford storage room for 3,000 or more bushels of sweet potatoes, and will keep them in excellent condition, if all details receive proper attention. The requirements for successful storage are that the tubers shall not be too hot, nor too cold, nor too wet, and that sudden changes of temperature shall be avoided. The sweet potato crop may be said to vary from 100 to 150 bushels per acre, under ordinary management, with higher results under good conditions.

CHAPTER V.

SASHES AND BEDDING PLANTS.

The cost of a hot-bed sash, glazed and painted, is somewhere about $2; and such a sash can be made to earn its cost every year. The farmer who has, say, a pair of sashes for hot-bed work and another pair for cold-frame work, can turn them to very good account in the early spring, not only in starting such bedding plants as may be required in his own operations, but in producing plants for his neighbors. It costs but little more to grow 1,000 than 100 cabbage, tomato or egg plants, and the surplus above the home requirement can be converted into dollars.

The Hot Bed.—The hot-bed is merely a board-lined pit, containing fermenting manure, with a few inches of soil on the manure, and covered by a sash. The ordinary sash is about 3 × 6 feet. A board shutter, the exact size of the sash, or a mat of straw, completes the outfit. The depth of manure, depending on the purpose in view, should be from 1 to 2 feet, the depth of soil from 3 to 6 inches, and the distance from soil to glass about 4 inches at the start. As the manure ferments the soil will sink.

The Cold Frame.—The cold frame is merely a piece of rich, mellow soil, enclosed by boards and covered with glass. There is no bottom heat of any kind, but it is a great deal warmer than the open soil, and serves a variety of purposes.

In the hot-bed, made in February or March (in the latitude of Philadelphia), all tender things may be started. The usual seeds sown here at that date on heat are cabbage, cauliflower, radish, lettuce, onions, etc., followed by tomato, pepper, celery, egg plant, etc., including flower seeds, if desired.

The cold frame is used through the winter for lettuce, onions, carrots, corn salad, spinach, etc., and in spring for the reception of the things started on heat, when the time arrives for transplanting and hardening them.

Properly-managed sashes will do a great deal toward the production of early market crops, and profits not infrequently depend upon the item of earliness.

The one thing for inexperienced persons to learn about sashes and their uses is the imperative necessity of free ventilation whenever the sun shines on the glass.

CHAPTER VI.

THE STRAWBERRY.

In addition to the several vegetables enumerated in the preceding pages, there is one of the small fruits that has taken such a prominent place in what may be termed farm horticulture as to deserve mention here. It is the strawberry.

This berry is, perhaps, the most popular small fruit in America, and because of its perishable character, is one that requires strictly local production. It cannot be shipped long distances without loss of character and flavor, and hence the local grower will never be crowded out of his own market.

The culture of the strawberry is simple and easy. There are many ways of setting out plants, and the after-treatment also differs widely. There will always be controversy concerning the respective merits of the hill system and the matted row system. Each cultivator must decide for himself which is the better.

For the farmer, whose acres are many and whose duties are pressing, there is, perhaps, no better way than to set strawberries in rows 4 feet apart, with plants 2 feet apart in the row, and to allow the plants to run together in the rows, giving sufficient attention to keep the alleys well stirred and the whole bed clear of weeds. To set an acre will require about 5,000 plants.

The winter covering of litter should be raked into the walks or alleys as soon as winter is over and allowed to remain there as a mulch for keeping the soil cool and damp and for the purpose of keeping the berries clean.

New Twice-Bearing French Strawberry "Mammoth Perpetual." For Description, see "Johnson & Stokes' Garden and Farm Manual."

As soon as the crop is off, the bed should be plowed, turning strawberries and litter under, and sweet corn or other quick crop at once planted. This will insure the gathering of two crops in two years; otherwise a strawberry crop means a two-years' use of the soil.

The setting out of a new strawberry bed every spring is good practice; and it is altogether advisable for farmers to occasionally introduce new varieties of strawberries on their farms, to replace old or enfeebled sorts.

The profits of strawberry culture are quite large, the gross receipts not infrequently running to $250 per acre. New boxes and crates are advisable, and are distinctly profitable.

IRRIGATION

BY CHEAP MODERN METHODS

Tells you just what you want to know in just the way you want to be told.

There is something here for every farmer and gardener. That thing is a sufficient water-supply. Irrigation makes deserts to rejoice and gardens to blossom. Nature often withholds needed moisture at critical times. Millions of dollars are lost annually through the uncertainties of the weather. After reading this work you will be surprised at the cheapness and practicability of irrigation, which will double the production at an expense of from $10 to $1,000, just according to what you want to spend, while it reduces soil culture to an exact science and enables the tiller of the soil to work on schedule time.

For terms on which this book can be had, see our "Garden and Farm Manual," which is sent free to all who write for it.

Floracroft Seed Gardens and Trial Grounds

In order to get the best results from our efforts, and make sure that customers shall receive from us the best seeds that the world produces, we have for many years maintained and carried on extensive trials at our Floracroft Seed Gardens and Trial Grounds, located about nine miles from our city warehouses. All operations are under the personal direction and management of one of our firm, who resides there. Here are planted each season, for thorough trial, samples of all "Novelties" offered by other seedsmen both in this country and Europe, as well as anything which may be sent us, claimed to be new and superior, by our amateur or market garden customers. By this means we are enabled to satisfy ourselves of the true character and value of any novelty before it can find a place in our catalogue.

Many acres are also devoted to the production of pedigree stock seed, from which the seeds we offer are grown. We plant the best seeds obtainable; then go over the crop, plant by plant, carefully "rogueing" and destroying the inferior and selecting and saving only the best. This stock seed from selected plants is sent to be grown on our farms in localities where the conditions of soil and climate are best adapted to the perfect development of the particular variety. It is the product of such stock seed only that we offer for sale.

Here are also located our Seed Testing Houses, where a sample of every lot of seed, whether grown by ourselves or grown for us under contract, is thoroughly tested, in mother earth, for vitality and purity of stock, and only those of satisfactory quality and germinating power are sold. In fact, we leave no stone unturned to gain and hold the confidence of all customers and secure them from disappointment.

Printed by BoD™ in Norderstedt, Germany